Low Temperature Plasma in the Processes of Functional Coating Preparation 2013

Journal of Physics: Conference Series Volume 479

Kazan, Russian Federation
4-7 November 2013

ISBN: 978-1-62993-941-4
ISSN: 1742-6588

Printed from e-media with permission by:

Curran Associates, Inc.
57 Morehouse Lane
Red Hook, NY 12571

proceedings.com

Some format issues inherent in the e-media version may also appear in this print version.

Printed by Curran Associates, Inc. (2014)

For permission requests, please contact the Institute of Physics
at the address below.

Institute of Physics
Dirac House, Temple Back
Bristol BS1 6BE UK

Phone: 44 1 17 929 7481
Fax: 44 1 17 920 0979

techtracking@iop.org

Additional copies of this publication are available from:

Curran Associates, Inc.
57 Morehouse Lane
Red Hook, NY 12571 USA
Phone: 845-758-0400
Fax: 845-758-2634
Email: curran@proceedings.com
Web: www.proceedings.com

TABLE OF CONTENTS

Journal of Physics: Conference Series **479** (2013) 012001 doi:10.1088/1742-6596/479/1/012001

Study of dust-electron plasma discharge on condition parallel of the velocity vector of the gas flow and electric field

G Dautov[1], I Dautov[1], I Fayrushin[2] and N Kashapov[2]

1 Department of General Physics. Kazan National Research Technical University after name A.N.Tupolev – KAI, Kazan, Russian Federation

2 Department of Technical Physics and Energetics. Kazan Federal University, Kazan, Russian Federation

E-mail: fairushin_ilnaz@mail.ru

Abstract. Investigated non-independent electrical discharge in the longitudinal flow of dust-electron plasma. Found that the charge transfer in the discharge gap is mainly free electrons and positively charged dust particles.

1. Introduction

Dust-electron plasma is formed in combustion of gaseous and liquid fuels and gas-thermal plasma coating of parts and products in the plasma chemical reactors , in the chambers of jet engines for liquid and solid fuels, in MHD generator channels [1]. A number of studies [2-6] constructed mathematical models to carry out numerical calculations of the parameters dust-electron plasma, taking into account the size of the physical properties of the microparticles and composition of the plasma gas. Having thermionic emission from the surface of the microparticles results in the formation of free electrons, the positive charge particles compensating provided nonquasineutrality in general such a low-temperature plasma. However , there is still no reliable experimental data showing that the charge composition of the plasma dust-electron provided mainly positively charged microparticles and free electrons.

2. Experiments and results

Studies conducted in the experimental installation schematically shown in Fig. 1.

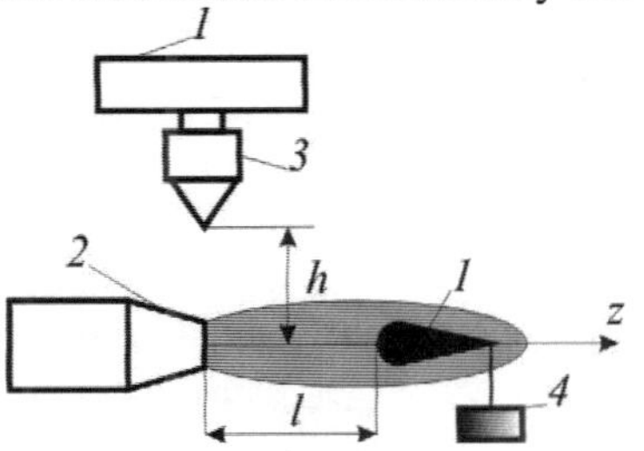

Fig. 1.

The distance between electrodes $l = 2\cdot10^{-2}$ m exhibited at moving the movable ceramic platform *4*. One electrode shape for reducing the droplet flow disturbances along the dust *z*. *2* served electrode nozzle of the gas burner . The output section of the dispenser *3* was located at a height h = $3\cdot10^{-2}$ m from the axis of the flow at the nozzle exit *2* , powder varied by controlling the input voltage electromagnetic vibrator *5* and maintained the value of $g = 4{,}8\cdot10^{-5}$kg / s . The average temperature of the plasma flow was varied from 1500 to 1900 K by adjusting the flow of oxygen in the burner , the flow rate of propane was maintained constant at a value of $2{,}4\cdot10^{-5}$ kg / s . In experiments used powders KCl and NaCl, consisting of particles with diameters of $4\cdot10^{-5}$ to $6{,}3\cdot10^{-5}$ m and the powder of Al_2O_3 - $5\cdot10^{-5}$ m embodiment using two input voltage , the electrode *1* - cathode (below the text "straight polarity"), one electrode - the anode ("reverse polarity").

The use of chemical compounds at a temperature of the plasma provides a significant thermionic emission from the surface of the dust particles and at the same time , due to the formation of ions , can get the values of the current non-self- discharge on the orders of her superior absolute error of measurement instruments applied in the transition areas. The particle size and the selected flow rate *g* provide great value of their total surface area and sufficient residence time in the discharge volume of the bulk of the dust component of the plasma. Were dust particles weighing KCl, NaCl and Al_2O_3 collected on the electrode 1 , on the movable platform 4 and scattered along the interelectrode gap at a distance $l = 2\cdot10^{-2}$m when they are in the plasma flow . A large percentage of particles reaches the electrode 1 and passes through the plasma discharge.

Fig . 2 shows the temperature distribution along the radius *r* of the plasma stream dust-electron at *z* = 1,8·10-2 m (curve *1*) and *z* = 5·10-2 m (curve *2*). The origin of the z-axis is selected in the nozzle outlet of the gas burner , curves 1 and 2 the values obtained for the discharge current $I = 3{,}5\cdot10^{-4}$A at " straight polarity ", the discharge voltage U = 400 V , powder $g = 4{,}8\cdot10^{-5}$ kg / s . On the resulting chart to select an area of the curve 1 width of $5\cdot10^{-3}$ m, in which there is a large temperature gradient flow. When you move along in the direction of the discharge nozzle remains the distribution and the width of the selected area is reduced to $2\cdot10^{-3}$m (curve *2*). The temperature and the nature of its distribution remain when the U and the polarity of the electrodes . Thus, the highlighted areas for measuring the characteristics of dust-electron plasma averaged temperature was selected at the expense of propane and oxygen.

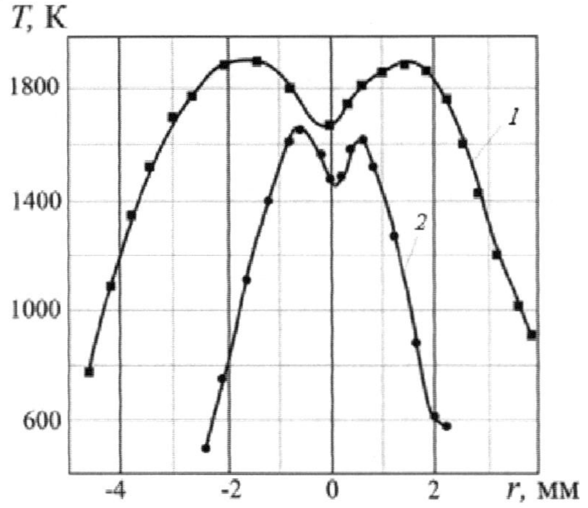

Fig. 2.

For a rough estimate of the contribution of various physical processes in the change of the current density used in the family of non-self-discharge voltage characteristics presented in figures 3-5.

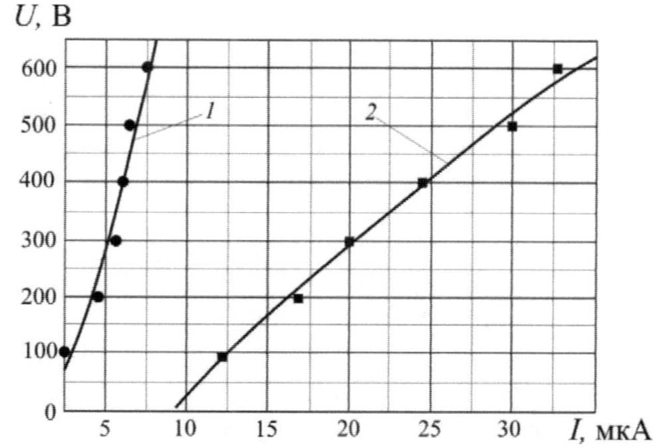

Fig. 3.

As can be seen from Figure 3, $g = 0$, $l = 2 \cdot 10$ 3m, $T = 1600$ K, a change of "reverse " direction (curve *1*) to " direct " (curve *2*) leads to an increase in the current. The value of the current in the "straight polarity " depends on how dirty the surface of the tungsten cathode sputtering products and duration of heating it in a stream of plasma , therefore , to assess the contribution of thermionic emission , we take the maximum difference CVC - $I = 25 \cdot 10^{-6}$A .

Passing through the plasma, dust particles can serve as sources of positive and negative ions. Assuming that the change of the polarity of the electrodes does not lead to significant changes in the conditions of formation of these ions , their contribution can be estimated by comparing the current-voltage characteristics obtained with reverse polarity combining bits at a rate of powder without it. Figure 4 curve *1* is obtained for g = 0 , curve *2* for $g = 4{,}8 \cdot 10^{-5}$kg / Al2O3 with reverse polarity and the same values of other parameters $l = 10 \cdot 2^{-3}$m, $T = 1600$ K , $h = 3 \cdot 10^{-2}$ m . In order not to take into account the possible contribution of the electrode processes for evaluation will be used for the entire period of maximum change in value of the difference *U* discharge CVC - $I = 2 \cdot 10^{-6}$A.

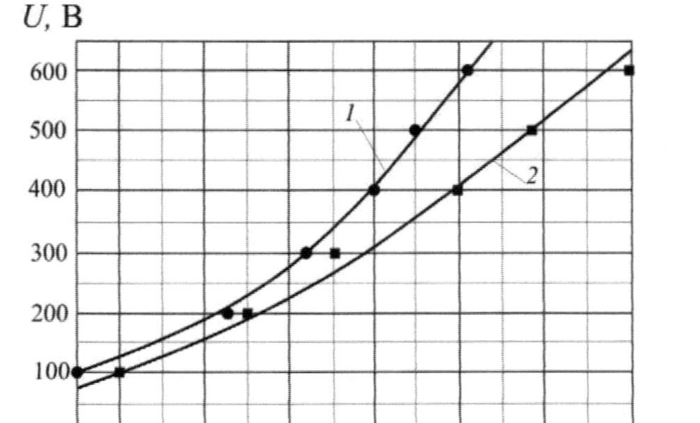

Fig. 4.

Shown in Fig. 5 CVC obtained with "normal polarity" and the values of the parameters $g = 4{,}8 \cdot 10^{-5}$ kg / s, $l = 10 \cdot 2^{-3}$m, $T = 1600$ K, $h = 3 \cdot 10^{-2}$ m, curves *1*,*2* and *3* of KCl, NaCl and Al_2O_3, respectively. The obtained values of the currents and the reduced estimate the contributions of different processes revealed that the charge transfer dust particles in coincidence directions of the flow rate pyleelektronnoy plasma and electric field non-self-discharge makes the greatest contribution to the high amperage.

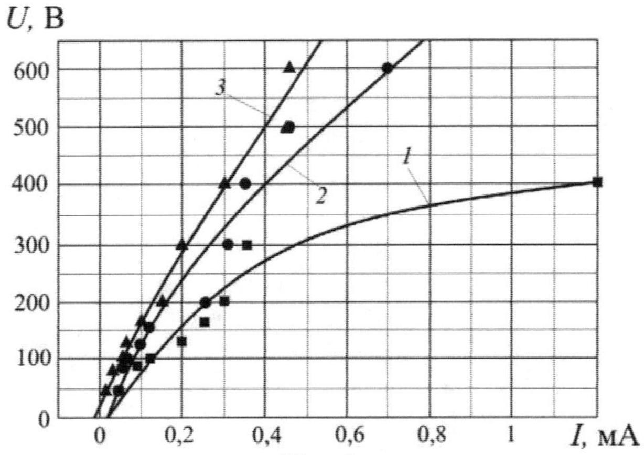

Fig. 5.

Unable to get a stable repeatability for KCl powder for discharge voltages $U \geq 400$ V. Depending on the environmental conditions (humidity and temperature of the air in the laboratory, the state of the cathode surface, etc.) when such discharge voltages change the current strength of up to several mA.

References

[1] Fortov V.E., Khrapak A.G., Yakubov I.T.Physics of strongly coupled plasmas. Textbook. allowance. M. FIZMATLIT , 2004 . 528 .
[2] Dautov G.Yu., Sabitov Sh.R., Fayrushin I.I. 2007 Vestnik of Tupolev KSTU **1** 29-32
[3] Dautov G.Yu., Dautov I.G., Fayrushin I.I. 2009 Vestnik of Tupolev KSTU **1** 57-59.
[4] Dautov I.G., Mardanshin R.M., Fayrushin I.I., Ashrapov T.F. 2010 Vestnik of Tupolev KSTU **3** 143-148.
[5] Dautov I.G., Kashapov N.F., Mardanshin RM, Fayrushin I.I. 2010 Vestnik of Tupolev KSTU **4** 134-136.
[6] V.I. Vishnyakov 2012 Phys. Rev. E **85** 026402.

Journal of Physics: Conference Series **479** (2013) 012002 doi:10.1088/1742-6596/479/1/012002

ION-PLASMA MODULES FOR APPLICATION OF NANOSTRUCTURED CARBON COATINGS

V T Barchenko[1], A A Lisenkov [1,2] and S A Trifonov[1]
[1]Sankt Petersburg State Electrotechnical University "LETI" Saint-Petersburg Russia
[2] Institute of Problems of Mechanical Engineering (IPME)
Russian Academy of Sciences (RAS) Saint-Petersburg Russia

s.trifonov.89@rambler.ru
Key words: vacuum system, magnetron sputtering, carbon thin films

Abstract. The basics of forming technology of nanostructured films in the form of composites containing hydrogenated carbon and metal parts was considered. Provides information about the methods used to generate these coatings. A description of the vacuum system provides the most satisfying modern requirements to technological equipment were given: minimum weight and size, power consumption, high stability and environmental safety of the process, tolerance to automate the process.

Thin carbon films in the amorphous state (α-C) containing bound hydrogen, are called hydrogenated films (α-C: H). The films are submit material with variable optical, electrical and mechanical properties during production process. This gives up great possibilities for application of these films in various devices [1]. In particular, nanostructured films of amorphous hydrogenated carbon with ferromagnetic nanoparticles are promising structures to formation of radio- absorbing materials (RAM) [2-4].

There are two classes of technologies of forming α-C and α-C: H films - physical vapor deposition (PVD) and chemical vapor deposition (CVD). The first group includes magnetron sputtering system, vacuum arc evaporators, as well as pulsed laser evaporation of graphite and a cathode sputtering with an ion beam of inert gas (in the case of α-C films). The methods of CVD include: ion-assisted deposition of hydrocarbon plasma, liquid phase epitaxy, pyrolysis, etc.

It is found that the properties of these coatings are highly dependent on the method and parameters of technological process of their formation, as well as the hydrogen concentration. This is determined by the type of chemical bonds between atoms in the film (a kind of hybridization of valence electrons: sp^2 or sp^3). The ratio of phases with different hybridization can be estimate by using energy-loss spectroscopy of electrons.

Systematic description of the methods of carbon deposition, the properties of the films and the technological features of their preparation may be found in [5]. Among the results of research in developing methods for producing films as ion-plasma and gas-phase methods received by domestic by the authors, it can be noted in [6] and [7].

Among the above-mentioned technologies for producing carbon films in many ways stands out the ion-plasma technology based on magnetron sputtering systems. Flow of plasma evaporated particles is generated in highly non-equilibrium conditions. This fact defines a very weak effect of temperature on the parameters of the sputtering process. This circumstance allows flexible regulation of the process parameters of the films and affects their physical and chemical properties and performance characteristics.

The information about the basics of technology of composite hydrogenized carbon thin films with metallic inclusions developed by the modernization of vacuum system ZV-1200 made by company LEYBOLD HERAEUS is given below.

This system is a gateway and a multimodule installation, which allow carrying process for several samples simultaneously. Location of modules installation is shown in Figure 1.

1 – Load chamber, 2- Provisional pump and the final purification chamber, 3- Technological chamber, 4 – Cooling chamber, 5 – Unloading chamber.

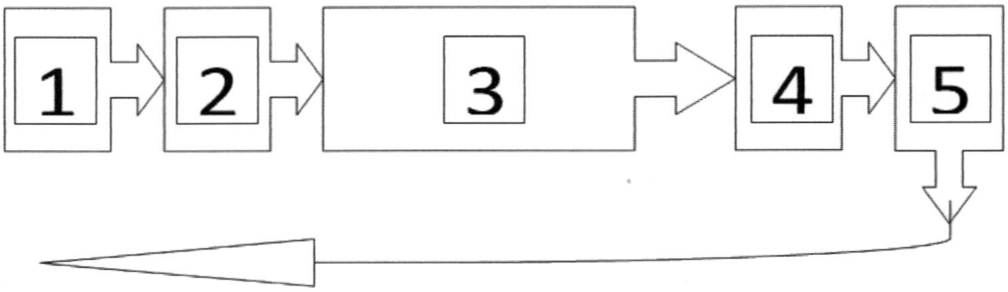

Figure 1 Block - diagram of the ZV-1200 vacuum system

Working algorithm of sputtering system can be summarized as follows. Loading of substrates is carried out in chamber 1. Next, the substrate is moved to chamber 2, which have dry pump system provided by turbomolecular pumps. There occurs degassing of products and finishing ion cleaning of surfaces, which are supposed to be covered by the film. Ultimate pressure in the working volume of the chamber can be up to 10^{-4} Pa.

To improve the quality and performance of products before the film formation process the substrate material must be subjected to finishing treatment to activate the surface and remove the water vapor and residual gases from the surface.

In the basic installation finish ion cleaning occurred in the glow discharge in argon at pressures in the range of 2 - 5 Pa. The energy of the ions bombarding the surface, determined by the automatic offset, which is formed to ensure the vanishing of the charge brought in from the plasma to isolated or dielectric surface. In the absence of a magnetic field near the surface the magnitude of the potential difference in the boundary layer is approximately $7kT_e$ (k - Boltzmann constant , T_e - the electron temperature) and for the conditions under consideration is in the range of 20 - 30 V. These energies of ions are close to the critical potential of sputtering, so cleaning conditions in the considered system are far from optimal.

To improve the quality of the finish ion cleaning was developed plasma source of fast neutrals on the basis of non-self glow, construction similar to that described in [8].

The main difference of the developed module is the fact that the energy disjoint-charge of the ions is less than 30 eV, which eliminates the appearance of the charge embedded in the surface layers of the structure and reduces heating of treated surface.

The second feature is the automatic stabilization of the surface potential, which can vary due to the charging of the surface due to the secondary electron emission during its bombardment by fast neutrals.

The process of film formation occurs in chamber 3. Design of this module is provided to ensure maximum coverage area with a given degree of non-uniformity and the possibility of the films application on heat sensitive materials. The module has the possibility to install up to 8 magnetrons (4 on each side) with geometric dimensions 750 * 80 mm disposed vertically. This modification decreases coating non-uniformity to 10% that was observed at the edges of the 600-mm long sample.

Film thickness uniformity in the horizontal direction is achieved by scanning of sheet of material mounted on intracameral element (pallet) with respect to the magnetron at a speed of 0.03 m/s. When applying single-component films, scanning amplitude decrease and deposition rate decrease in the chamber are achieved by using of parallel work of the magnetrons.

Using the targets made of dissimilar materials and independent regulation of the power supplied to the magnetrons allows sputtering composition coatings with controlled properties.

The maximum temperature to which the surface of material is heated during sputtering process is regulating by the selection of magnetron power and a scan rate of the material.

After cooling inside chamber 4, the sample is unloaded to the chamber 5.

Control part of the sputtering system had undergone deep modernization. The system is controlled by means of a specially developed software for PCs in the programming language C + +. There are both manual and fully automatic modes of operation of the sputtering system.

In automatic mode, the parameters of the process are supported with a high accuracy by providing feedback to the functional sites, whose modes alter the properties of the formed films.

The basic installation was implemented the ability to handle 2 samples simultaneously. This upgrade is done because of impossibility to handle large amount of data on the basis of outdated electronic components. Therefore, to improve the efficiency of all modules and enhance the performance of sputtering system currently being upgraded software based on programmable industrial data acquisition system and control of processing equipment. Final transfer of the installation to the modern standard of data exchange via RS- 485 interface is implemented by replacing the central controller. It will integrate additional equipment, such as components of magnetron sputtering system and means of mass spectrometry with the possibility of feedback with the main device without significant changes of software architecture.

References

[1] Konshina E A 2010 *Amorphous hydrogenised carbon in optical devices* (S-Petersburg: NRU ITMO)

[2] Nikolaichuk G A, Yakovlev S V, Trifonov S A and others 2008 Hydrogenized carbon sputtered Ultra wide band multilayer radio-absorbing coatings with 3d-metal magnetic nanoparticles *The 18th International Crimean Conference "Microwave & Telecommunication Technology"* **2** 579-580

[3] Nikolaychuk G A, Petrov V V, Yakovlev S V and others 2009 Radio-absorbing materials on the base of nanostructures *Nanotekhnika* **1 (17)** 41-44

[4] Yakovlev S V, Ivanov V P, Nikolaichuk G A 2009 Complex dielectric permittivity measurement of thin-film radio-absorbing materials in 300 MHz to 40 GHz frequency range using the

resonator method *Proceedings of the XVII International Conference "Magnetism, near and far spin-spin interaction."* S 57-58

[5] Robertson J 2002 Diamond-like amorphous carbon *Materials Science and Engineering* R **37** 129-281

[6] Oskomov K V 2001 Ion-plasma methods of applying solid amorphous carbon coatings on large area substrates *Dissertation for the PH.D. Institute of High Current Electronics (IHCE)* Tomsk 18

[7] Zolotukhin A A, Obraztcov A N, Ustinov S A, Volkov A P 2003 Formation of nano-carbon thin film materials in gas-discharge plasma *ZhETF* **124(6)** 1291

[8] Barchenko V T, Bystrov Y A, Kolgin E A, ed. Bystrov Y A 2001 *Ion-plasma technologies in electronic production* (St. Petersburg Branch: Energoatomizdat)

Investigation of the influence of plasma-electrolytic processing on the surface of austenitic chromium-nickel steels

L N Kashapov[1], N F Kashapov[1] and R N Kashapov[1,2]
[1] Institute of Physics, Kazan Federal University, Kazan, 420008, Russia
[2] Laboratory of Radiation Physics, Kazan Physical-Technical Institute of the Kazan Scientific Center of the Russian Academy of Sciences, Kazan, 420029, Russia

E-mail: kashramiln@gmail.com

Abstract. Investigated the plasma-electrolytic processing of austenitic chromium-nickel steel. Determined that the treatment process consists of three parts: the anodic dissolution of metal anode microdischarge area burned and the area in which the combustion gas discharge - Glow vapor-gas layer around the electrode without the occurrence of certain microdischarge. The effect of processing parameters on the morphology of the surface.

1. Introduction

Today it is the austenitic steel in the world the most widely used, the percentage of use is 60 - 70% of total steel consumption. One of the main consumers of austenitic steels is the chemical industry. The necessity of the effective processing exists because from this type of steels produced most of containers, vessels, reactors and other equipment. However, the currently available methods of electrophysical and electrochemical processing have drawbacks - the difficulty of obtaining the required microrelief, high energy consumption, environmental contamination. Solution of this problem can be found in applying of the new method – plasma-electrolytic processing. This method of processing is carried out by applying a voltage to the electrodes which are in a liquid electrolyte in result is the electrolyte plasma in a thin layer near the active electrode. [1]. Discharge burns in the combined-cycle shell between the metal electrode and the electrolyte surface. Depending on the type of the processed electrode plasma-electrolytic process is divided into the anodic and cathodic. There is a clearing and polishing of the workpiece during the cathodic process due to the burning of the gas discharge, are also possible options for the application of coatings under certain conditions. When the anodic process in the according from the kind of metal is either anodic oxidation with microarcs combustion, the process is called microarc oxidation [2] or anodic dissolution with his erosive destruction [3].

The aim of this study was to determine the basic parameters of the anodic process of plasma-electrolytic processing of austenitic steels.

2. Experimental

To achieve this objective an experimental setup was used [4]. In this paper investigated the influence of ripple voltage. Investigations were carried out on samples of 12X18H9T (12% - C, 18% - Cr, 9% - Ni, Ti < 1%). To prevent changes in the size of the area of the anode putted on the dielectric protective case which leaves open only a well-defined part of the electrode. Anode area was 2.9 cm^2 area of the cathode – 18,61 cm^2., S_k=6,4*S_a. Electrolytic bath was filled with electrolyte - sodium chloride aqueous solution of concentration: 1%, 3% and 5% by weight. For each concentration of the solution was carried out removal of the current-voltage characteristics of the plasma-electrolytic process. Processing of the samples was carried out at values of no-load voltage ~ 45, 90, 130, 180, 220 V. The main operating parameters of the electrode system are: the voltage on the electrodes, the magnitude of the discharge current, the current density on the electrodes. Changing the form of current and voltage at the time of discharge ignition was determined by the oscilloscope FLUKE 105 SCOPEMETER SERIES II, timebase varied from 5 ns to 60 s. Measuring the voltage on the anode was carried out with a digital universal measuring device APPA 105 relative measurement error of 0.1%. With a digital multimeter APPA 305 we measured the temperature of electrolytic cathode with copel-chromel thermocouple. The electrolyte temperature before and after of the experiment measured by mercury thermometer. To determine the average linear speed of metal removal be weighed sample before (m_1) and after (m_2) processing. And then was calculated by the following formula:

$$\omega = \frac{\Delta m}{\rho \cdot S_a \cdot \Delta t}$$

where $\Delta m = m_1 - m_2$; ρ - density of steel 12X18H9T equal to 7.92 10^3 kg/m3 S_a - the area of the anode; Δt - the processing time.

The surface morphology of the samples was obtained after the plasma-electrolytic processing and was visualized with a scanning electron microscope Carl ZEISS EVO 50. Also were taken a profilograms of the surface by profilograph-profilometerTR-200. Were calculated values of the height of the surface roughness in the ten points R_z the arithmetic average roughness R_a, the maximum height profile R_y, depth of maximum depression profile R_m, mean square roughness R_q, the total height of the irregularities R_t, the average pitch unevenness of the profile S_m.

3. Results

Volt-ampere characteristics (VAC) of the anodic process of plasma-electrolytic processing of steel 12X18H9T in solutions with different concentrations is shown in Figure 1.

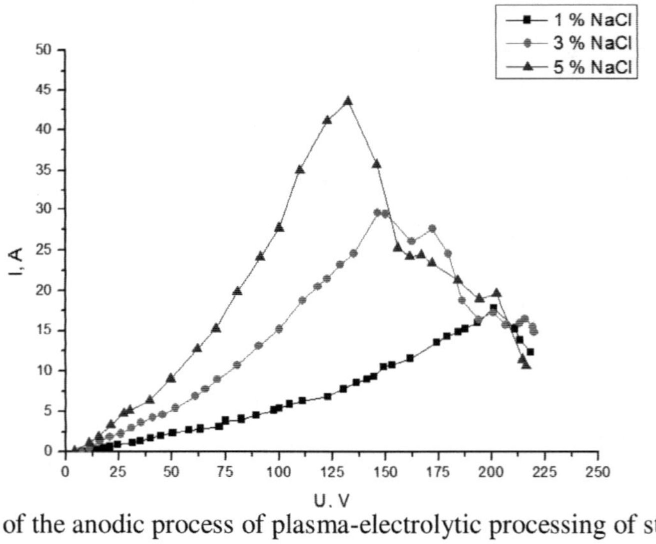

Figure 1. VAC of the anodic process of plasma-electrolytic processing of steel 12X18H9T

All VACs have a maximum, the increase in concentration of the solution leads to a shift of the peak to the area of smaller voltages and increase from 15 to 45 A. To solution with a concentration 1% the voltage of initiation of anodic microdischarges was ~ 140 V, while a concentration of 3 and 5% ~ 90 - 100 V. You may also notice that, for voltages above 200 V all of three graphs are the same, and the coincidence of concentrations for 3 and 5% occurs at 150 V. Current-voltage characteristics can be divided into three areas: area of the electrolysis - anodic dissolution, burning of anodic microdischarges and the third area - the appearance of plasma formation around the anode, at which the discharge is produced between the metal anode and electrolytic cathode.

Offset maximum of VAC to the area of smaller voltages due to the increase of the specific conductivity, which depends on the solution concentration and temperature. Measuring the average temperature of the electrolyte showed that during the receiving VAC process average temperature of the electrolyte for a concentration of 1% change from 20 to 75 ° C, 3% - to 88 ° C., 5% - to 88 ° C. Thus, it is possible to reject the effect of temperature on the VAC offset relative to each other and leave a major factor in the solution concentration.

The second area is burning of anodic microdischarges together with anodic dissolution. As can be seen from the curves of VAC the concentration of the solution has a significant effect, namely the initiation of anodic microdischarges with increasing concentrations occur at lower voltages.

The third area - the appearance of the plasma formation is accompanied by a decrease of the current and corresponds to the maximum point of the curve VAC. Since the VACs in the area for different concentrations is the same, we can say that the concentration of sodium chloride does not affect on the discharge burning between the metal anode and the cathode electrolytic. Consider this process in terms of influence on the surface of austenitic chromium-nickel steels. On the basis of current-voltage characteristics were selected modes of processing steel. Models of processing and parameters of roughness shown in Table 1.

Table 1. Modes plasma electrolytic processing, and surface roughness

Conc. of solution	U without temperature (B)		t (sec.)	U_{mean}	I_{mean}	Δm (mg)	ω mm/s	R_a micron	R_z micron	R_y micron	R_m micron	R_q micron	R_t micron	S_m (mm)
1 %	A	45	47,1	35	1,8	25	0,231	3,549	12,66	22,16	14,29	4,672	27,9	0,129
	B	90	30,5	82,3	4	40	0,571	4,337	12,35	26,69	15,93	5,79	38,79	0,2157
	C	130	31,3	119	7,5	86	1,188	6,775	20,28	39,11	22,42	8,657	51,4	0,1725
	D	180	30,6	165	8	79	1,124	4,137	16,39	25,06	14,74	5,176	32,85	0,1161
	E	220	30,3	205	22	73	1,048	4,017	16,8	23,21	14,17	4,979	27,51	0,09
3 %	A	45	30,6	40	4,3	38	0,540	3,088	11,13	16,92	10,41	3,872	22,34	0,1291
	B	90	30,6	80	9	91	1,294	2,782	6,66	17,43	8,552	3,658	26,84	0,2834
	C	130	30,6	115	15	104	1,479	2,409	10,28	19,71	13,3	3,474	26,18	0,1236
	D	180	30,9	155	19	88	1,239	3,027	12,39	19,8	12,35	3,894	25,2	0,1042
	E	220	30	200	25	93	1,349	1,785	8,29	12,9	8,57	2,371	17,01	0,085
5 %	A	45	30,3	39	8	78	1,120	0,74	2,63	4,13	2,076	0,919	5,45	0,1103
	B	90	30,5	75	24	178	2,540	5,553	21,3	33,53	18,59	7,053	44,07	0,1141
	C	130	30,4	107	27	145	2,076	2,871	12,03	18,63	11,62	3,732	22,88	0,0896
	D	180	30,5	135	30	129	1,841	1,329	6,38	9,5	5,67	1,723	14,07	0,0566
	E	220	30,7	185	25	96	1,361	0,805	4,45	6,11	3,98	1,025	7,34	0,0978
-	-		-	-	-	-	-	0,409	1,993	3,207	2,151	0,550	4,519	0,0634

Since during the treatment process there is a change voltage and current values in the table are represented by their average values.

Metal removal rate depends nonlinearly on the applied voltage at first increasing and then going down. Increasing the concentration of the solution leads to a shift the maximum to the area of smaller voltages and growth, so for maximum concentration of 1% and 3% adequate mode "C", and the 5% - "B". The highest removal rate was 2.54 10^{-3} mm / sec for the mode "B".

All five processing modes can be divided into groups based on the influence on processed surface, look at an example a 3% solution. First group - is modes A and B. The surfaces of A and B are similar in appearance, difference is a large the difference of the roughness parameters and the linear speed of metal removal. This can be attributed by big power: Mode A - 176 Watt, V - 720 Watt. Thus in modes B and C there are same kind of physical impact, the only measure of the influence in the second case more. The second group is the modes of processing C and D. Both samples across the surface of coated with microwells size less than 1 micron. Also on the sample for the mode C observed microwells larger size from 10 to 40 microns, and for the D - microwells larger size from 10 to 50 microns. Measurement of roughness parameters showed that there is an increase R_z, R_a, R_y. The third group is the mode E. The surface of the sample processed under these conditions has a "sponge-like" structure, all surface is coated with microwell size is less than 1 micron, in some places they are superposed on each other forming microdefects size of 2-3 microns.

Changing the concentration is changed border processing mode. To a solution of 1% concentration of sodium chloride of the third group was not observed. First group - A and B, the second group - C, D, E. And for a solution with 5% concentration of NaCl is observed quite a different picture. First group – A, second group – B and C, and the third group - D, E.

Conclusion:
Research of plasma-electrolytic processing that it is an effective method to obtain required value of microrelief. Erosion the destruction of occurs as a result of combustion a large number of separate microdischarges between the metal anode and the cathode in the electrolytic gas bubbles formed at the electrode surface at different times and developing parallel to each other.

It is established that the entire process is divided into three parts: the anodic dissolution of the metal, burning of anodic microdischarges and area in the which the combustion gas discharge - the glow steam gas layer around the electrode without causing the individual microsparks. By setting the voltage corresponding to one of three treatment regimens can achieve the desired surface characteristics.

Research on the influence of change of concentration of electrolytes in the plasma-electrolytic processing have shown that increasing the concentration effect on the anodic dissolution of the metal and the occurrence of anodic microdischarges, but does not affect the discharge in the steam-gas layer. The results of plasma-electrolytic formation of the microstructure of the surface of austenitic chromium-nickel steels have shown that the precision machining, surface finish, the depth of the defect layer can be adjusted by varying the processing mode in the final stages of formation.

References
[1] Kashapov R.N., Kashapov N.F. 2005 4 International Symposium on Theoretical and Applied Plasma Chemistry. 361-364.
[2] Suminov I.V., Epelfeld A.V., Lyudin V.B., Crete B.L., Borisov A.M. Microarc oxidation (the theory, technology, equipment Textbook. 2005 368
[3] Kashapov R.N. 2008 Advanced Materials **5** 466-469.
[4] Kashapov R.N. 2010 Physics and Chemistry of Materials Processing **5** 50-56.

Journal of Physics: Conference Series **479** (2013) 012004 doi:10.1088/1742-6596/479/1/012004

Combustion instability control in the model of combustion chamber

AN Akhmadullin, EN Ahmethanov, OV Iovleva, GA Mitrofanov

Kazan (Volga) Federal University

E-mail: techphyskfu@gmail.ru

Abstract. An experimental study of the influence of external periodic perturbations on the instability of the combustion chamber in a pulsating combustion. As an external periodic disturbances were used sound waves emitted by the electrodynamics. The purpose of the study was to determine the possibility of using the method of external periodic perturbation to control the combustion instability. The study was conducted on a specially created model of the combustion chamber with a swirl burner in the frequency range from 100 to 1400 Hz. The study found that the method of external periodic perturbations may be used to control combustion instability. Depending on the frequency of the external periodic perturbation is observed as an increase and decrease in the amplitude of the oscillations in the combustion chamber. These effects are due to the mechanisms of synchronous and asynchronous action. External periodic disturbance generated in the path feeding the gaseous fuel, showing the high efficiency of the method of management in terms of energy costs. Power required to initiate periodic disturbances (50 W) is significantly smaller than the thermal capacity of the combustion chamber (100 kW).

1. Introduction

The increase in thermal stress of work processes in the combustion chambers of power plants is accompanied by oscillatory processes (pressure fluctuations, temperature, speed, etc.) [1]. They emerge and develop spontaneously. Oscillation frequency depends on the geometry of the combustion chamber. The amplitude of the oscillations depends on the degree of coherence of pulsations of heat in the combustion zone with pressure fluctuations (Rayleigh criterion) [2] , gas mass flow , vortex shedding and other time-dependent processes. With pulsating combustion increases substantially complete combustion of fuel, heat transfer to the heating surface, marked increase in the rate of combustion. Thus, the use of pulsating combustion mode can substantially increase the efficiency of combustion processes.

In real pulsating combustion furnace devices, as the operating mode of the combustion chamber is not provided. This is due to the negative effects that accompany the pulsating combustion. First of all, this powerful sound waves that can reach 150 dB. In addition, the continued existence of a pulsating mode of combustion -chamber organization requires a special volume. This requires the use of special burners, make a concerted acoustic characteristics of the feed fuel lines, combustion chambers and pipes, which discharge the products of combustion, etc.

Currently, control of combustion instability using a number of methods [3] , including the method of external periodic perturbations. As an external periodic perturbation can be used sound waves emitted by, for example, electrodynamics. The control action is performed on the supply path of gaseous propellants or directly into the combustion chamber.

An experimental study was conducted to determine the feasibility of using the method of external periodic perturbation to control the instability of the combustion chamber in a pulsating combustion.

2. The method of investigation

Schematic of the experimental chamber is shown in Figure 1.

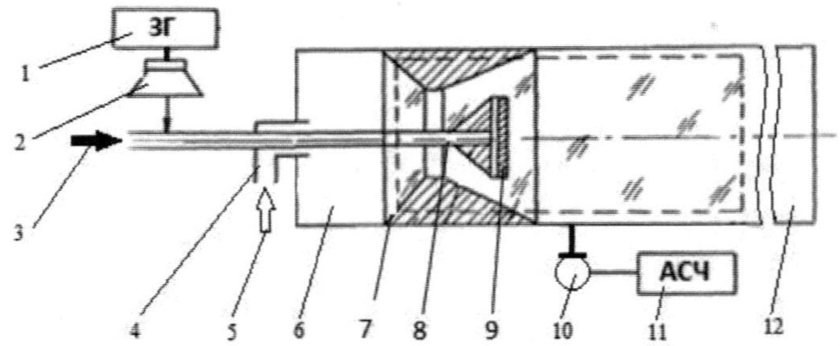

Figure 1. pulsating combustion chamber.

The combustion chamber (CC) consists of the supply of gaseous fuels tract 8 , the air circuit 4, the air tank 6, 7 diffuser, 9 of the flame , the resonance tube 12. Sound vibrations created by sound generator 1 and fed into the fuel supply path through electrodynamics 2. Pressure fluctuations in the combustion chamber detected by the microphone 10 and the frequency of the spectral analyzer 11. GHG created by burning propane mixture. Gaseous fuels reported by the anti- flame. The stabilizer had a chance to move in the cavity of the diffuser. This changes the conditions for vortex air flow regulator. When the optimum flow velocity v of the air flow regulator in the chamber was excited pulsating combustion. Sound pressure oscillations in the combustion chamber reaches 130-150 dB depending on the fuel consumption.The result of the influence of external periodic disturbances estimated by the change of the existence of pulsating combustion in the combustion chamber (Fig. 2, 3).

Construction of the existence regions of pulsating combustion was carried out in the following sequence. Asked a certain flow rate of gas and air in the combustion chamber. For the obtained ratio of fuel components made tests the excitation pulse combustion in the range of possible movements of the flame. The harmonic signal generator created a certain amplitude sound waves and fed on electrodynamics. When moving in the cavity of the flame cone and the fixed coordinates of the excitation termination of pulsating combustion in the combustion chamber . The boundaries of the regions of existence of intermittent burning in the coordinate system of thermal stress of the combustion chamber $\boldsymbol{q}$ and the flow rate at the edges of the flame $\boldsymbol{v}$ determined by calculation. Studies carried out in the frequency range from 100 to 1400 Hz.

3. Results and their discussion

The amplitude and frequency of the periodic disturbance entering the combustion chamber, depending on the amplitude -frequency characteristics of the feed path of gaseous fuel and the combustion chamber. The combustor is a complex cavity in which both exhibit the properties as a half-wave and quarter-wave resonators. As the frequency of the external periodic disturbances in the range of resonance frequencies of the combustion chamber there is the effect of synchronization with increasing amplitude of sound pressure oscillations. On the disturbance frequencies close to resonance are observed beats, which depending on the amplitude of the disturbance can disrupt pulsating combustion within the combustion chamber. At frequencies of periodic disturbances antiresonant frequencies close to the combustion chamber, there is the effect of suppressing oscillation. At the heart of these events are synchronous and asynchronous mechanisms of action [3].

Most clearly the effect of external periodic perturbations on the combustion instability manifests itself through a change of the existence of areas of intermittent burning. Figure 2 shows the existence of pulsating combustion for different sound pressure levels of external periodic perturbations proportional to the electrical power to the electrodynamic ***We***. The existence of regions constructed in the coordinate space of the combustion chamber thermal tensions q_S and flow of gaseous fuel to the edge of the flame ***v***. The length of the combustion chamber and the resonator was 138 cm frequency disturbance corresponded to the first antiresonant frequency of the combustion chamber as a half-wave resonator.

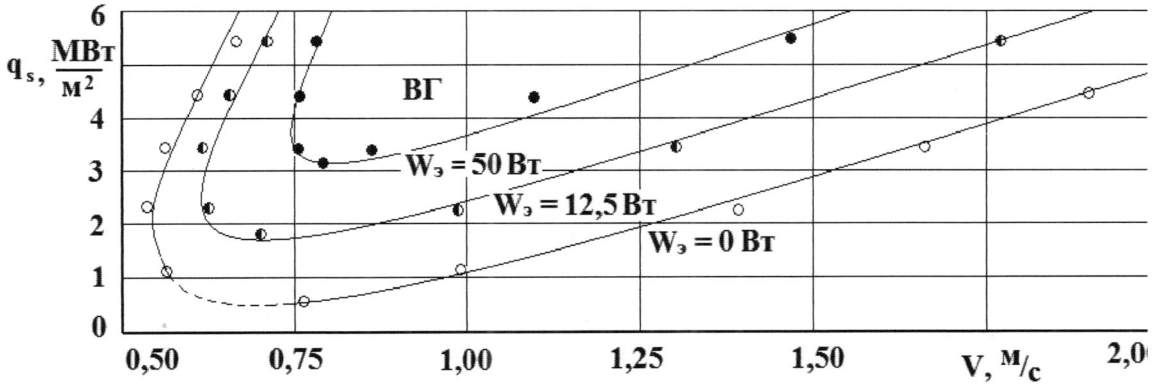

Figure 2. Changing the scope of the existence of intermittent burning (frequency perturbations generated 180 Hz)

The increase in electric power invested in the creation of periodic acoustic disturbances on electrodynamics, leads to a significant reduction of the existence of pulsating combustion. During the pilot study discovered the effect of " high- impact" external periodic disturbances on the oscillations in the combustion chamber. This effect is evident in the frequency range from 870 to 1320 Hz (Fig. 3)

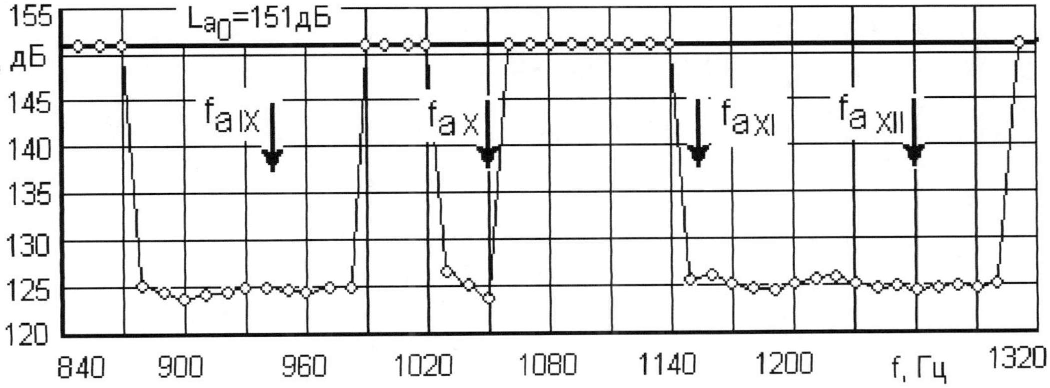

Fig. 3 . Graphs of the resulting sound level the pressure in the combustion chamber

The length of the combustion chamber was 138 cm , the heat intensity of the combustion chamber $q_S = 1{,}1$ MW/m^2. The initial level of sound pressure oscillations intrachamber was 151 dB. As a result of the perturbation level of sound pressure oscillations decreased milking 125 dB. The suppression of oscillation in the range IX - XII calculated resonant modes of the combustion chamber (the first mode oscillation is 105 Hz) . At the same time consumed electric power was 10.5 watts. Figure 4 shows the existence of pulsating combustion in the case of high- impact.

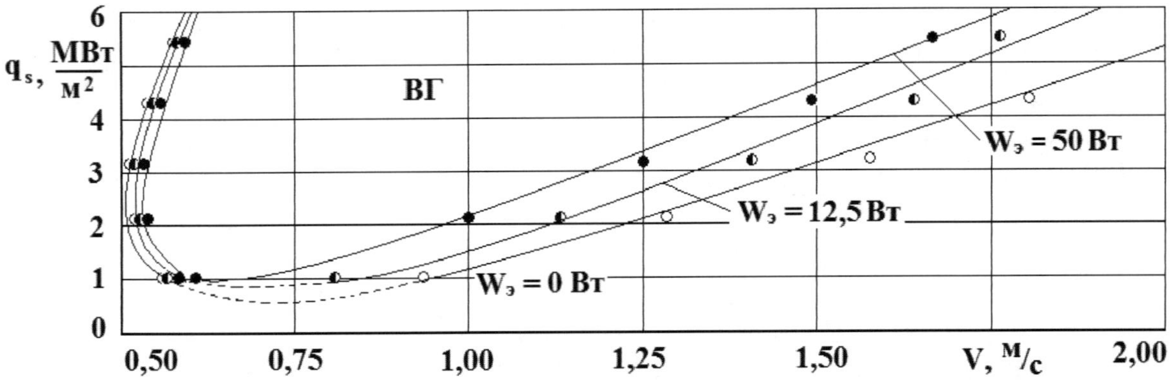

Figure 4 . Changing the scope of the existence of intermittent burning (The frequency of disturbances created in 1210 Hz)

The phenomenon of " high- impact" presumably due to the sensitivity of the gas jets sound vibrations . Sound vibrations lead to destruction of actively laminar and turbulent jets for active mixing of the external environment . In the case of combustion - the intensification of the process by reducing the preparation time of fuel to combustion.

4. Conclusions

- Method of external periodic perturbations may be used for controlling combustion instability in the combustion chambers of power plants highly stressed.
- Depending on the frequency of the external periodic perturbation is observed as an increase and decrease in the amplitude of the oscillations in the combustion chamber. These effects are due to synchronous and asynchronous mechanisms of action [3]
- External periodic disturbance generated in the path feeding the gaseous fuel , showing the high efficiency of the method of management in terms of energy costs. Power required to initiate periodic disturbances significantly less heat output of the combustion chamber.

References

[1] Friedland S V, Tazmeev A Kh, Miftakhov M N and Tazmeev Kh K 2006 *Vestnik mashinostroeniya* **7** 72
[2] Babichev A P, Babushkina N A, Bratkovskij A M et al 1991 *Fizicheskie velichiny: Spravochnik* (Moscow: Energo atom izdat)

Research of plasma-electrolytic processes in various ratio of the area anode to the cathode

L N Kashapov[1], N F Kashapov[1] and R N Kashapov[1,2]
[1] Institute of Physics, Kazan Federal University, Kazan, 420008, Russia
[2] Laboratory of Radiation Physics, Kazan Physical-Technical Institute of the Kazan Scientific Center of the Russian Academy of Sciences, Kazan, 420029, Russia

E-mail: kashramiln@gmail.com

Abstract. The object of research is the plasma-electrolytic processing, namely the influence of area ratio of the anode and the cathode to the parameters of the process. Defined area ratio anode to cathode, at which the discharge burning occurs on the cathode, on the anode or on the both electrodes simultaneously. Determine the required ratio of the electrodes ($S_1 \geq 2\ S_2$) for discharge burning of the electrode S_2

1. Introduction

Recently, most research efforts aimed at the study and application of low-temperature plasma produced when using liquid electrodes. Gas discharge is initiated on the surface of one of the electrodes immersed in an electrolyte, wherein burning takes place between solid and liquid. Given combination of electrodes allows the polishing, cleaning, formation of micro-relief surface [1,2], the application of functional coatings [3], diagnosis and cleaning fluid [4]. Despite the numerous practical applications of this variety of gas discharge at the moment there is no consensus about its nature and its main parameters. It is considered that the discharge occurs subject to inequality of surface areas of solid electrodes $S_1 > 5\ S_2$ [5,6], immersed in an electrolyte. It is assumed that the discharge burns in the steam shell, which was formed by the heating liquid around the electrode. But it is also possibility to ignition of the discharge in gas bubbles formed on the surface of the electrode [7]. At present there is no information on which of the electrodes – on the anode or on the cathode will be the initiation of a gas discharge when they have the same areas, and depending on the type of electrolyte used. Therefore, the aim of this work is to determine the influence of ratio of the areas electrodes on the course of plasma-electrolytic processing.

2. Experimental

The study was performed on the installation, the functional diagram is shown in Figure 1. It consists of a current source 1, electrode system 2, the oscilloscope 3, series resistor 4, of the plating bath 5, voltmeter 6, thermocouple 7, ammeter 8.

The electrode system is formed so that the electrodes are vertically fixed by horizontal metal bolts, through which current flows. To prevent changes in the size of the area on the electrodes were putted on the dielectric protective covers that are left open just a well-defined part of the electrode. Maximum surface area of electrode in contact with the surface of the electrolyte was 339.5 mm^2. For ignition discharge on any of electrodes its contacting surface decreased relative to initially a predetermined maximum value.

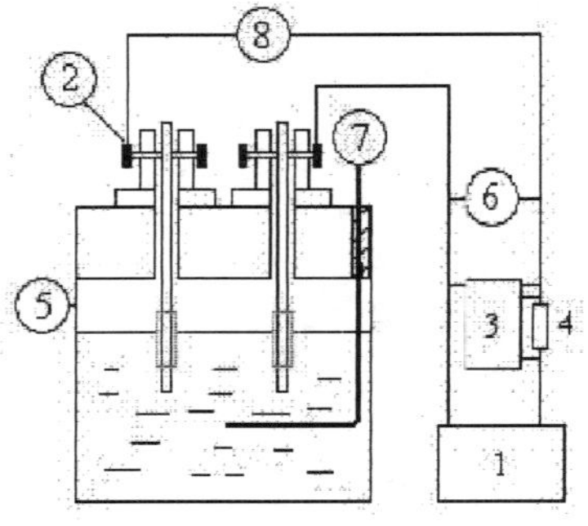

Figure 1. Functional diagram of the experimental setup.

Investigations were carried out on specimens made of steel grades 12X18H9T, Art. 3, copper, graphite. Before the experiment the samples were subjected to a thorough grinding and subsequent polishing. The electrolytic bath is filled with the studied electrolytes the required concentration and composition. As studied electrolytic cathode, an aqueous solution of $NaHCO_3$ 5% of concentration by weight, pH = 7.

The main operating parameters of the electrode system: the voltage on the electrodes, the magnitude of the discharge current, the current density on the electrodes. Voltage increased stepwise at an interval of 5 - 10 V. Measuring the voltage on the anode was carried out by a digital multimeter device APPA 105 relative measurement error of 0.1%. With a digital multimeter APPA 305 we measured current. Changing the shape of current and voltage at the time of discharge ignition was determined by the oscilloscope FLUKE 105 time base varied from 5 ns to 60 s.

3. Results

Let's consider initiation of a gas discharge in a solution of $NaHCO_3$ with different electrodes. For convenience of description, plasma-electrolytic process, introduce the parameter a / k, indicating the ratio of the area of the anode to the cathode area. When using electrodes made from steel 3, a ratio a / k = 1 discharge burned at the anode surface, the ignition voltage of 110 V. In the case of the other electrodes at voltages of 110 - 120 V short bursts were observed on the cathode, but with increasing voltage at the cathode chuffing disappeared and sustained discharge burning is initiated at the anode. In case of using steel electrodes 12X18H9T (0,12% - C, 18% - Cr, 9% - Ni and Ti < 1%) with a/k = 2,26 discharge began to burn on the cathode at a voltage of 110 V, while burning electrode is red-hot and began to melt. The current-voltage characteristics (CVC) plasma-electrolytic processing using steel electrodes 12X18H9T shown in Figure 2.

The working solution during the plasma-electrolytic processing with steel electrodes did not change. When using copper electrodes solution turns blue, preserving transparency.

Figure 3 shows current-voltage characteristics of a plasma-electrolytic processing using 5% solution of $NaHCO_3$, for various area ratios electrodes 3 made of steel, copper. Current-voltage characteristics of the discharge on the anode are different from the CVC discharge on the cathode by having a maximum. For CVC of the anodic process is characterized by a high near 130 V for nearly all electrodes, except for copper. In case of using copper electrodes at low voltages is a maximum due to the formation of the oxide layer and its subsequent destruction. The presence of this process leads to a shift of the main peak in the region of 150 V, and discharge initiation occurs at 120 V. In the case of the cathode of the process current-voltage characteristic increases linearly, then reaching a certain value and remains constant. Can distinguish that the discharge burning on the cathode occurs at a high current density than anode. Let us consider in more detail the current-voltage characteristics of plasma- electrolytic processing with electrodes made of steel 3. In an applied voltage of 170 V for the ratio a/k = 1 current density was 26 mA/mm^2, and for a/k = 1,76 - 52 mA/mm^2. Investigation of the ignition discharge at various a / k showed that it is possible to achieve the ratio at which the discharge will occur at both electrodes. For electrodes made from steel 3 in a ratio a/k = 1,45 alternately the discharge burning observed at both electrodes. But it is also burned in the cathode first, and then further increasing the voltage on the anode ignited. For the ratio a/k = 1,76 and above the discharge burning occurred only at the cathode. A similar pattern was observed for the copper electrodes. The figure clearly shows if you do not take into account the values of the currents that shape the curve for a/k = 1,45 is from itself an intermediate state between the curves for a/k = 1,76 and a/k = 1. And it is precisely this curve traced two maxima at 130 V and 170 V. When comparing the CVC for a/k = 1 and a/k = 1,76 with CVC for a/k = 1,45 shows that the peaks at 130 V are the same, difference is that the second peak, which arises from the dominance of the cathode discharge at high voltage. This behavior of the CVC is due to, that at different times the discharge burning prevails at any one of the electrodes.

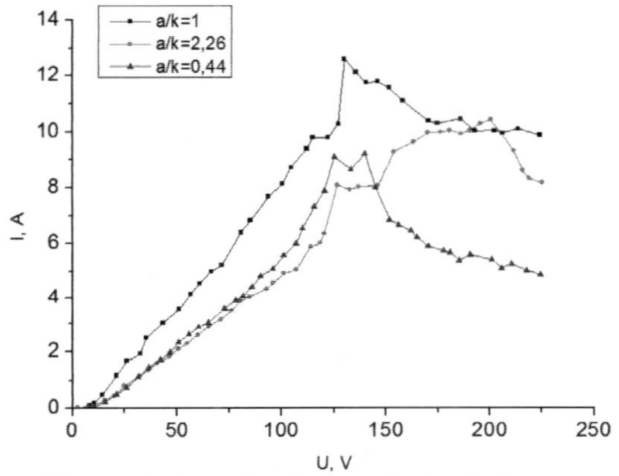

Figure 2. Current-voltage characteristics of a plasma-electrolytic processing using a 5% solution $NaHCO_3$

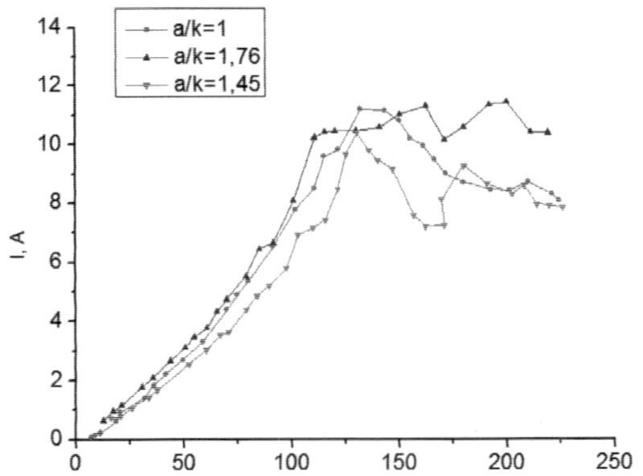

Figure 3a. Current-voltage characteristics of a plasma-electrolytic process using 5% $NaHCO_3$ solution with electrodes made steel 0,3%- C

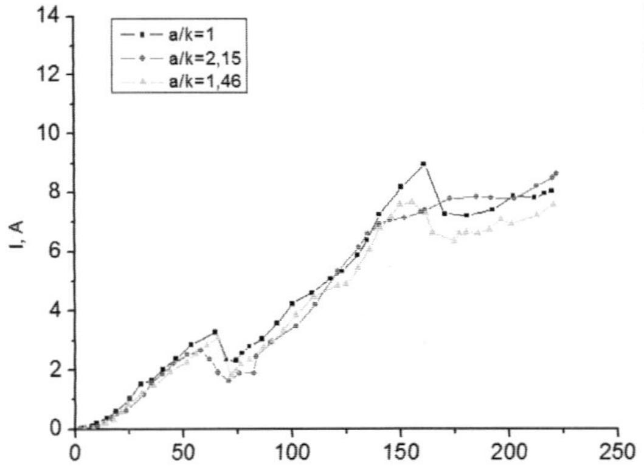

Figure 3b. Current-voltage characteristics of a plasma-electrolytic process using 5% $NaHCO_3$ solution with electrodes made copper

Using the ripple voltage allowed simultaneously initiates discharges both the anode and cathode. The applied voltage obtained after full wave rectification affects each new period. Analysis of the oscillograms confirms this statement. By selecting a certain area ratio anode to cathode (a / k = 1.46) allowed a variable discharge burning on the anode and cathode. Thus, a prerequisite for discharge burning of the electrode S_2 is the following relationship: $S_1 \geq 2\ S_2$. This will optimize the technological process of plasma-electrolytic processing and make it more energetically favorable.

Acknowledgments
The authors gratefully acknowledge the financial support from the Russian Foundation for Basic Research №13-08-00906 A.

References

[1] Kashapov R.N. 2010 Physics and Chemistry of Materials Processing **5** 50-56.
[2] Kashapov R.N. 2011 Herald KSTU **4** 149-154.
[3] Kashapov R.N., Kashapov N.F. 2005 4 International Symposium on Theoretical and Applied Plasma Chemistry. 361-364.
[4] Jin X.L., Wang X.Y., Yue J.J., Cai Y., Zhang H.G. 2010 Electrochimica Acta **56** 925-928
[5] Slovetskii D.I., Terent'ev S.D. 2003 High Energy Chemistry **37** 355.
[6] Slovetskii D.I., Terentev S.D. 2005 4 International Symposium on Theoretical and Applied Plasma Chemistry. 145-150.
[7] Suminov I.V., Epelfeld A.V., Lyudin V.B., Crete B.L., Borisov A.M. Microarc oxidation (the theory, technology, equipment Textbook. 2005 368

Journal of Physics: Conference Series **479** (2013) 012006 doi:10.1088/1742-6596/479/1/012006

On the separation of arc discharge plasma near the sputtering cathode

N A Pankin, N A Smolanov

Ogarev Mordovia State University, 430005, Saransk, Bolshevistskaya St. 68, Russia

E-mail: smolanovna@yandex.ru

Abstract. In the method of the molecular dynamics simulation of the motion of atoms and ions of titanium in spatial inhomogeneous electric and magnetic fields. An attempt is made to explain the experimental distribution of the coating thickness on the surface of the condensed ring anode installation NNV, at the film condensation method with ion bombardment.

1. Introduction

In many papers [1-4] on the calculation of the distribution of the coating thickness when applied ion- plasma methods do not include the effect of an external magnetic field (even if it is provided condensation technology). In this case, the trajectory of the charged particles is taken close to linear. This is true in low magnetic fields (up to several hundred gauss) and a relatively high ion energies (from tens of eV). At the same time, within the framework of these models can not explain the distribution of the thickness of the coating produced near the cathode (Figure 1) [5]. The experimental scheme is shown on Figure 2 (1 - cathode, 2 - substrate, 3 , 4 - stabilizing and focusing coils, respectively, O - origin; L - distance cathode -substrate).

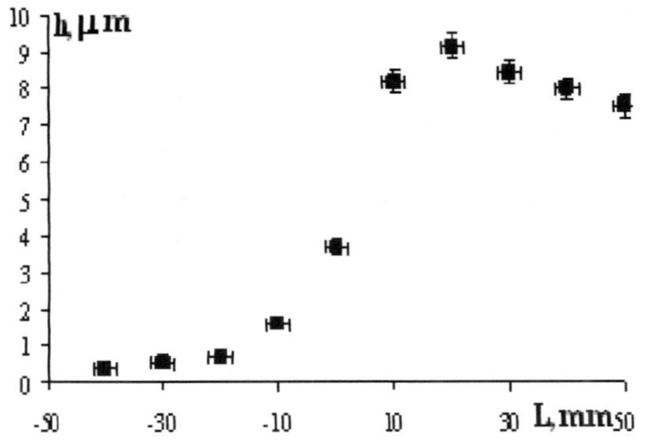

Figure. 1. Distribution of film thickness

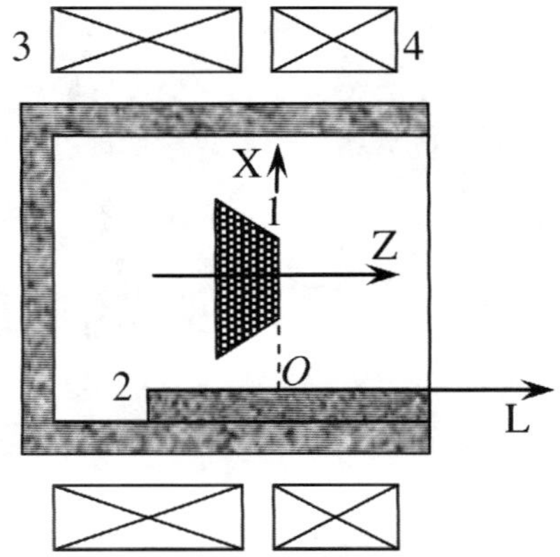

Figure. 2. The scheme condensation coating

Journal of Physics: Conference Series **479** (2013) 012006 doi:10.1088/1742-6596/479/1/012006

The aim of this work was the theoretical calculation of the coating thickness depends on the geometry of the samples near the placement of titanium cathode. The calculation scheme is shown in Figure 2. Thus treated, the titanium ions having a charge of from 0 to 3*e* (*e* - the charge of an electron). Taken into account the external magnetic field using the system of focusing and stabilizing coils position NNV and the bias potential. The ion energy, the process simulation, varied from 1 to 100 eV.

2. The model calculation

The simulation was performed using the method of molecular dynamics [6]. The equation of motion of each of the ions and atoms of titanium can be written as:

$$m_i \frac{d^2\vec{r}_i}{dt^2} = q_i\left[\vec{v}_i\vec{B}\right] + q_i\vec{E},$$

where r_i - radius vector of the particle ; m_i - mass of the particle ; q_i - the charge of the particle ; v_i - the ion velocity , B , E - the magnetic field strength of the electric field, respectively . When building the model does not take into account the interaction of ions with one another. To solve the above system of differential equations of the second order Verle algorithm is used in the form of high-speed [6] . The time step is set equal to 10 ns. In the calculations it was assumed that the distribution of particles in the directions of their departure in the emission from the cathode surface is described by the cosφ (φ - angle measured from the normal).

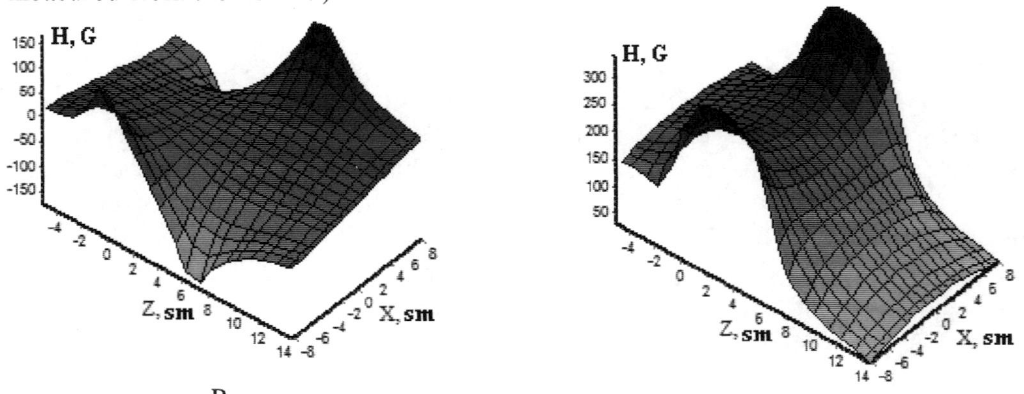

B_X B_Z

Figure. 3. The distribution of the magnetic field components

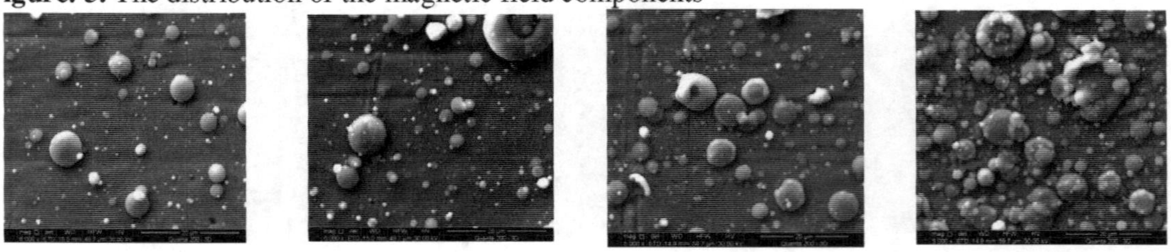

а) L=-2 sm *б) L=-1 sm* *в) L=0 sm* *г) L=1 sm*

Figure. 4. The surface coating of different areas of the anode ring

3. Results and discussion

The results of calculating the distribution of the magnetic field for the system shown in Figure 3. The geometrical parameters and operating modes of the simulated system consistent modes of operations and geometric parameters set NNV-6.6I4 in the coating by condensation with ion bombardment. [7]

The simulation results of the ions of different charge show that when a negative potential difference leads to stretching of the area near the cathode, low energy ions (10 eV). Increase and decrease of charging energy emitted from the cathode of ions leads to the possibility of condensation on the surface of the annular anode. In the simulation we have not found the condensation of atoms and ions of titanium for $L< 0$. When the experimental results measuring the thickness of X-ray methods (Fig. 1) and electron microscopy (Fig. 4) indicate the presence of condensed matter at distances of L is less than zero.

The possibility of condensation in this area is associated with the collision and recombination processes in ion-plasma flow of low-temperature plasma. In this area, condensation may only atomic particles and high-energy ions reflected from the surface of the particulates that are present in the vacuum-arc deposition of films (Figure 4) [5, 7-10]. Also, creating a stream of particles moving in the range of negative values of L may occur in the Coulomb explosion of charged polyatomic structures (clusters, etc.) , the atom - atom collisions , the emission of particles , "drip" spot , etc. Thus, the basis of the atomic flux particles are particles as ions are strongly pulled out of the area near the cathode by the magnetic and electric fields. Furthermore, a substantial contribution to the thickness distribution of the condensate on the surface of annular anode making drip temperature plasma component, the content of which increases with the distance L.

References

[1] Markov G V 1996 *Physics and Chemistry of Materials Processing.* **3** 71 (in Russian)
[2] Karatunov Yu V 2003 *Diss. PhD* (Moskva: MSIET) (in Russian)
[3] Poluektov N P 2004 *Diss. PhD.* (Moskva: MSIET) (in Russian)
[4] Koba V I, Lunev I V, Padalka V G 1992 *Physics and Chemistry of Materials Processing.* **2** 94 (in Russian)
[5] Pankin N A, Smolanov N A 2006 *Journal of Surface Investigation. X-ray, Synchrotron and Neutron Techniques.* **10** 54 (in Russian)
[6] Heerman D V 1990 *Computer Simulation Methods in Theoretical Physics* (Moskva: Science) (in Russian)
[7] Aksenov I I, Andreev A A, Bren V G 1979 *Ukrainian Journal of Physics.* **24(4)** 515 (in Russian)
[8] Shiao M H, Shieu F S 2001 *Thin Solid Films.* **386** 27.
[9] Horoshih V.M. 2004 *PES.* **2(4)** 200 (in Russian)
[10] Anders A 2008 *Cathodic Arcs: From Fractal Spots to Energetic Condensation* (Berkeley: Springer Science, Business Media, LLC)

Journal of Physics: Conference Series **479** (2013) 012007 doi:10.1088/1742-6596/479/1/012007

Ion-plasma way of receiving strengthening heat reflecting anti-reflection coating

R.T. Galyautdinov, N.F. Kashapov

Department of Technical Physics and Energetics. Kazan Federal University, Kazan, Russian Federation

E-mail: raf7g@mail.ru

Abstract. The ion-plasma deposition strengthening heat reflecting clarifying coatings is developed for plastic screens of means of individual protection. The method of magnetron sputtering are received strengthening coatings with integrated coefficient of a transmission in visible area of a range more than 96%.

1. Introduction

Acceleration of scientific and technical progress is accompanied by increase of risk of technogenic accidents at which there is an environmental pollution by highly toxic substances. Pollution of the soil, water and air causes significant economic damage. Elimination of consequences requires existence of the rescue services which employees have to be reliably protected from influence of harmful substances. The most well-tried remedies of protection at elimination of technogenic accidents are isolating suits which help with the prevention and elimination of consequences of technogenic accidents and natural cataclysms. In the isolating suits intended for work in harmful conditions, transparent plastic filters are applied.

Plastic screens made now insufficiently strong and in the conditions of hostile environment lose over time transparency. For prevention of it it is necessary to put a sheeting on an external surface of plastic filters. The available means of individual protection (MIP) have a short resource of work because the filters, a part MIP, lose the properties under the influence of hostile environment. For increase in service life of protective suits production of the filters possessing a high resource of work in the conditions of aggressive environment is actual.

For hardening of plastic screens ion-plasma technologies are perspective. Plasma methods of drawing optical coverings [1-5] allow to strengthen the MIP plastic filters for work of staff of rescue services at elimination of consequences of technogenic accidents and natural cataclysms. The purpose of work is development of technology of plasma drawing strengthening heat reflecting anti-reflection coating for the MIP plastic screens

2. Equipment and Methods

The unique UVN-70A-2 [3] stand was developed for deposition strengthening heat reflecting anti-reflection coating, with the automated complex of registration of parameters of plasma. The schematic diagram of installation of a high vacuum magnetron sputtering coating machine is given in figure 1. The following techniques and measuring equipment were applied to pilot studies of properties of the received strengthening coverings and their dependence on parameters of the abnormal smoldering category in the crossed electric and magnetic fields: measurements of spatial distribution of floating potential of electric field by a probe method, measurement of a magnetic field by a magnetic probe, temperature distribution by the hromel-kopelevy thermocouple, voltmeter of a direct current, ampermeter of a direct current, Hitachi 330 spectrophotometer, Phillips XL30 ESEM TMP scanning electronic microscope (SEM).

The system of pumping of the modernized vacuum UVN-70A-2 installation provides residual pressure in the vacuum camera of $6{,}5 \cdot 10^{-3}$ Pa and the subsequent operation of the magnetron with a pressure of working gas from 0,1 to 0,4 Pa. Into system of pumping enter: the pump vacuum lamellar and rotor NVR-16D, the pump vacuum diffusive N-400, a trap water-cooled. Deposition coatings is carried out by the magnetron sputtering system consisting of two identical magnetrons with replaceable cathodes therefore there is an opportunity to put difficult multilayered coverings in one running cycle. The system of gas supply includes cylinders with plasma-forming gases. Optical thickness of put layers are supervised by system of photometric control on change of intensity passing through a put coating and a transparent substrate of a ray of light.

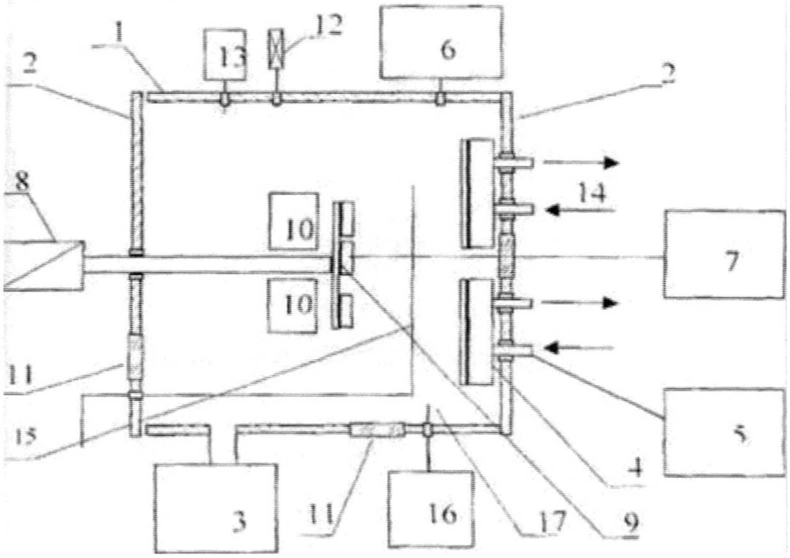

Fig. 1 - Schematic diagram of high vacuum magnetron sputtering coating machine: 1 - vacuum camera; 2 - lateral covers; 3 - system of vacuum pumping; 4 - magnetrons; 5 - power supply; 6 - gas supply system; 7 - system of photometric control of thickness; 8 - system of rotation of substrates; 9 - processed details; 10 - heaters; 11 - observation ports; 12 - valve of start-up of air; 13 - manometrical sensor; 14 - the union of supply of cooling water; 15 - gate; 16 - high-voltage block of ionic cleaning; 17-high-voltage electrode of ionic cleaning

In the process of deposition coating the vacuum and partsialny pressure of residual gases, temperature of substrates and evaporators, speed of rotation of substrates, tension on electrodes, digit current, specific capacity on a target, an induction of a magnetic field are controlled.

Key parameters of the discharge changed in the following limits: tension on the cathode U changed from 375 to 600 V, category P_p capacity from 0,6 to 6 KV, distance from a magnetron l surface to a substrate from 0,1 to 0,2 m, pressure of working gas in a chamber p was supported from 0,1 to 0,3 Pa, as a material of the cathode served copper, the titan and stainless steel, time of a dusting of t changed from 0,5 to 25 min., the size of an induction of a magnetic field In changed discretely and made 0,02, 0,04 and 0,08 T, as plasma-forming gas served argon (consumption of gas $4{,}3 \cdot 10\text{-}4 \div 8{,}8 \cdot 10^{-7}$ kg / c) and an argon mix (consumption of gas $4{,}0 \cdot 10\text{-}4 \div 6{,}3 \cdot 10^{-7}$ kg/ c) with oxygen (consumption of gas $1{,}3 \cdot 10\text{-}4 \div 1{,}5 \cdot 10^{-7}$ kg/ c). Degree of a sparseness of a gas stream is defined by Knudsen's (Kn) number. For process of a dusting non-uniform on structure of coverings of Kn $\approx 0{,}1 \div 0{,}3$ that corresponds to a molecular current of gas.

3. Results

Production of a coating carry out in the vacuum camera of the modernized vacuum installation UVN-70A-2. Formation on a transparent plastic product of a strengthening adhesive layer of oxide of the silicon, a heat reflecting layer of oxide of stannum and a clarifying layer of dioxide of silicon carry out magnetron sputtering in the environment of argon and oxygen. Before the room of a transparent plastic product in the vacuum camera it previously are cleared.

In the vacuum camera with a residual pressure of 2,6 10-3 Pa carry out an argon blousing up to the pressure of 0,2-0,3 Pa, then a transparent plastic product close the gate and light the discharge on the magnetron with a target from silicon for removal of an oxidic film from a target surface within 5 minutes of burning of the discharge then add oxygen up to the pressure of 0,1-0,3 Pa, further the gate clean and deposit on a transparent plastic substrate strengthening adhesive layer of oxide of silicon SiOx with an optical thickness of (2,7 - 3,3)*λ0/4, where λ0 = 550 nanometers. At a smaller thickness of a strengthening adhesive layer of oxide of silicon strength characteristics of a covering sharply decrease, at a bigger thickness optical properties worsen. On the end of formation on a transparent plastic substrate of a strengthening adhesive underlayer of oxide of silicon light the discharge on the magnetron with a target from stannum and on a surface of a strengthening adhesive underlayer of oxide of silicon form heat reflecting layer of oxide of stannum SnO2 with an optical thickness of (3,6 - 4,4)*λ0/4. At a smaller thickness of a layer of oxide of tin the reflection coefficient in infrared area sharply decreases, at a bigger thickness optical properties worsen. On the end of formation of a heat reflecting layer of oxide of stannum light the discharge on the magnetron with a target from silicon and on a surface of heat reflecting oxide of stannum of a layer form clarifying layer of dioxide of SiO2 silicon with an optical thickness of (0,9 - 1,1)*λ0/4. At thickness other than range optical properties of a coating worsen.

In fig. 2 the strengthening heat reflecting anti-reflection coating on the plastic substrate, received is schematically presented in the declared way: 1 – substrate (transparent plastic); 2 – strengthening adhesive layer of oxide of SiO_x silicon ($1,5 \leq x < 2,0$); 3 – heat reflecting layer of oxide of SnO_2 tin; 4 – anti-reflection coating of dioxide of SiO_2 silicon.

In fig. 3 the range of a transmission of the strengthening heatreflecting anti-reflection coating received in the declared way, in visible area of a range is presented.

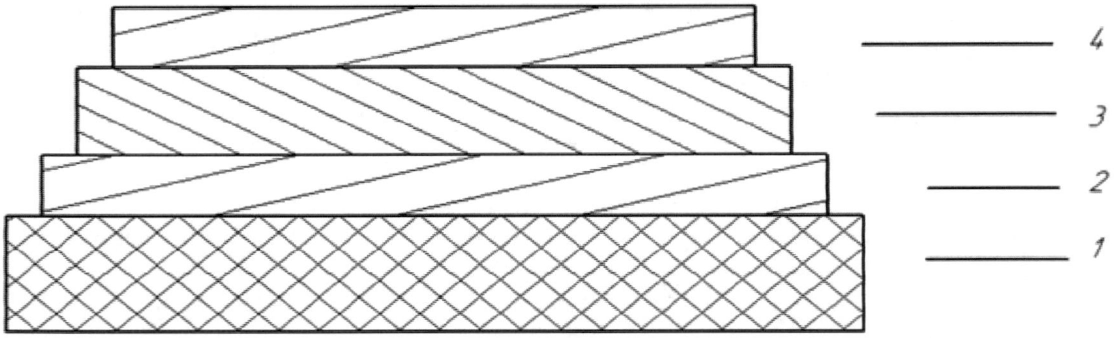

Fig. 2 Design of a strengthening heat reflecting anti-reflection coating

The method of magnetron sputtering are received strengthening heat reflecting protective coatings on polymeric substrates

The coating possesses the highest 0-th group of mechanical durability on an abrasion on OST 3–1901–95 for transparent plastic products (maintains more than 3000 turns on SM-55), integrated coefficient of a transmission in visible area of a range (450-650 nanometers) more than 96% (figs. 3)

and reflection coefficient in visible area of a range (450-650 nanometers) less than 2%, reflection coefficient in infrared area of a range more than 80%. Moisture fastness of coverings corresponds to 1 group on OST 3–1901–95 (maintain influence of humidity of 98% at a temperature (98 + 2)^{0}C within 10 days) and possess firmness in relation to chemically hostile environment.

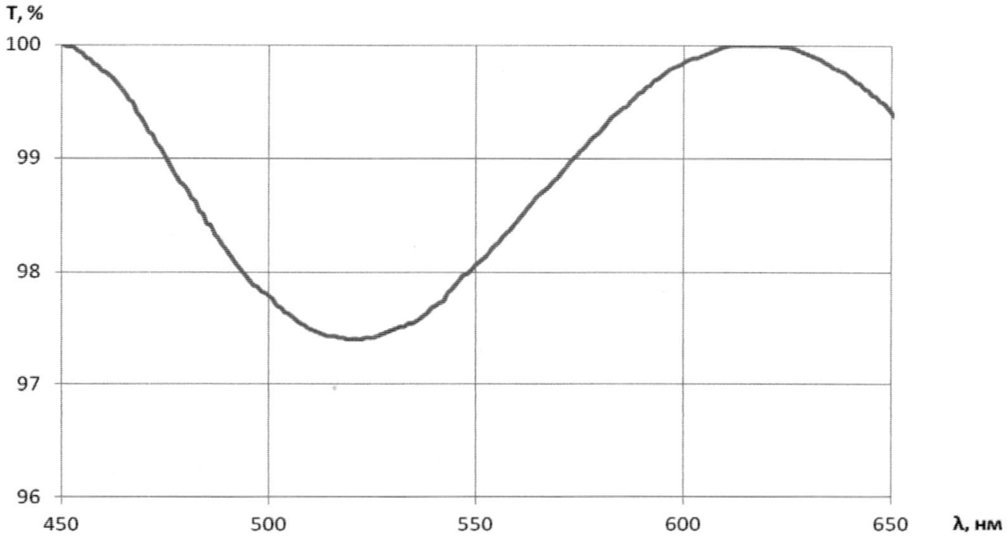

Fig. 3 Spectral coefficient of transmission of a strengthening heat reflecting anti-reflection coating

4.Conclusion

The ion-plasma deposition techniques of multilayered functional coverings are developed for organic substrates, including strengthening heat reflecting anti-reflection coatings for transparent plastic products, for example, for screens of means of individual protection.

References

[1] Galyautdinov R.T., Kashapov N.F., Luchkin G.S. 2005 Applied physics **6** 88-93.
[2] Galyautdinov A.R., Galyautdinov R.T., Kashapov N.F. 2007 Applied physics **6** 108-111.
[3] Galyautdinov R.T., Kashapov N.F. ,Tagirov R.F. 2011 Vestnik KSTU **10** 49-53.
[4] Galyautdinov R.T., Kashapov N.F. 2011 Vestnik KSTU **10** 58-63.
[5] Galyautdinov R.T., Luchkin G.S., Kashapov N.F. 2003 *Welding International.* **17**(8) 655-658.

Journal of Physics: Conference Series **479** (2013) 012008 doi:10.1088/1742-6596/479/1/012008

Ions neutralization by reflection from solid target surface

V T Barchenko and N A Babinov

Saint Petersburg Electrotechnical University, Saint Petersburg, Russia

E-mail: vtbarchenko@yandex.ru

Abstract. The paper focuses on the one of the ion neutralization processes – process of ion neutralization by reflection from solid target surface. The three-electrode system of the fast neutrals source is considered as well as the main factors determining the parameters of neutrals beam. Moreover, the loss of energy during neutralization was obtained and the graphs of energy loss versus incident angle of the ion to the target for some target materials were built.

1. Introduction

Formation of submicron size structures is the essential part of the micro- and nanoelectronics products creation. Currently the most common method for this is plasma-chemical or ion-beam etching. But there is number of problems appearing when using this technology for etching of dielectrics or high-resistance semiconductors. Firstly the target acquires surface charge causing etching speed decreasing or even termination of etching. Secondly the UV radiation presented in the plasma spectrum may have negative effect on the target.

To prevent surface charging slow electrons insertion (current compensation) or high-frequency power supply can be used. Ultraviolet irradiation of the target can be reduced via appropriate design of the device.

However, all of these lead to significant rise of technological equipment price. Moreover the paths of ions in beam distorted due to target's surface charge and beam's space charge that hinders uniform etching and thus do not allow further reduction the size of formed structure parts.

These problems can be resolved through using broad beam sours of fast neutrals. Neutral beam do not distorted due to any charges or magnet fields that may be present in technological equipment. Moreover plasma source of fast neutrals can use direct current power supply.

2. Ion beam neutralization

2.1. Neutralization efficiency

There are two main processes than can be used to effectively neutralize ion beam: the resonant recharge on the gas target and neutralization by reflection from solid target surface.

In this paper the process of ion neutralization by reflection from solid target surface is considered. The main advantage over the resonant recharge is its independence of the pressure in the region of the neutralization. While the efficiency of the resonant recharge depends on the gas pressure the pressure in the sources of fast neutrals using ion neutralization by reflection from solid target surface is limited only by the pressure at which the discharge is maintained in plasma generation region and can be lowered down to 10^{-4} Torr using discharge with electrostatic confinement of electrons.

The process of neutralization by reflection from solid target surface is effective at low angles between the incident ion and the target. In [1] was simulated the passage of ion beam through the channel width of 0.1 cm, consisting of two plates, inclined at the certain angle to the transmitted beam and the dependence of the neutralization efficiency of the channel length and incident angel was obtained. (figure 1)

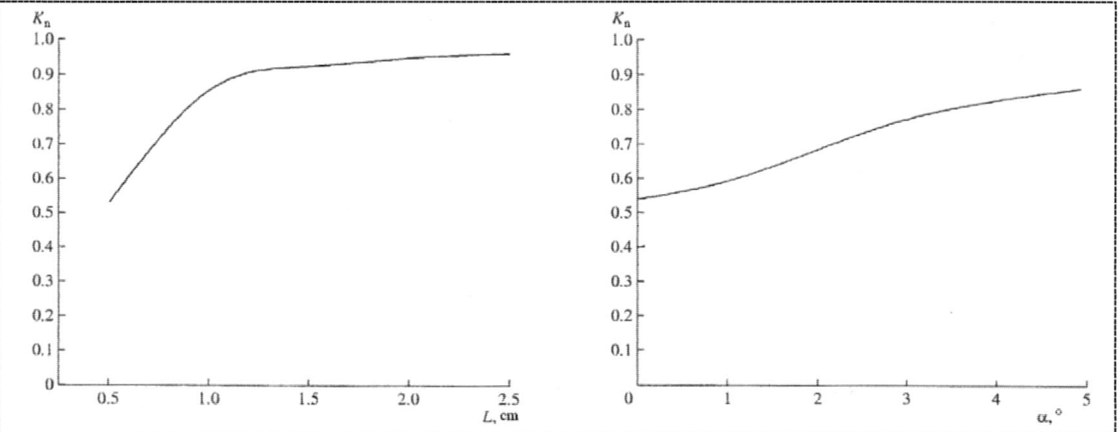

Figure 1. Dependence of the neutralization efficiency.

With sufficient length of the channel and low angle (angles everywhere reports from the tangent to the surface of the target) to the target surface coefficient of neutralization is close to unity and thus such a system effectively performs neutralization. But part of the ions in the beam still is passing through the channel without charge neutralization and fall onto the target surface. It leads to charge accumulation on the target surface and thus to charging of the surface. To avoid it the three-electrode system can be used (figure 2).

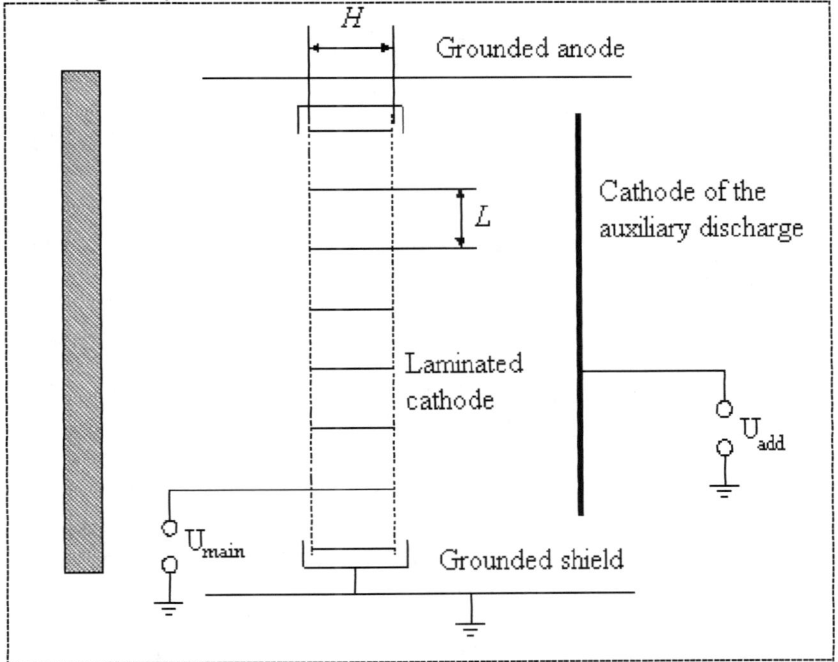

Figure 2. Three-electrode system.

Ions accelerated in the cathode potential drop are neutralized by reflection from the laminated cathode surface. At the same time not-recharged ions from the beam cannot leave potential well in the cathode region with sufficient energy and fall onto the target surface.

In such a system the distance between the plates is largely determines the characteristics of the system and cannot be less than two width of the cathode layer. Otherwise the plasma will enter into the gap between the plates and laminated cathode will function as the hollow cathode.

If the distance between the cathode plates is significantly less that its thickness we can ignore the hit of the ions on the end surface of the plates. In this case, we can approximately assume that the plasma forms a cylindrical surface around the end of the each plate with the radius of curvature equal to the width of the cathode layer (figure 3).

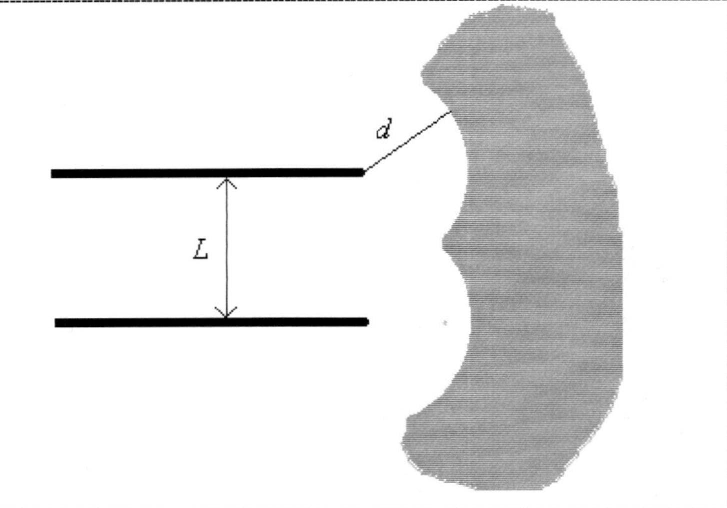

Figure 3. The location of the cathode plasma.

If we assume that ions are emitted normal to the surface of the plasma the maximum angle at which they fall onto the surface of the cathode equal to the $\arctan(\frac{L}{2d})$, where L is the distance between the plates of the cathode and d is the width of cathode layer. If we assume that ions neutralized by reflection when they fall to the surface at angles less than 10^{c} then the ratio of L to d should not be less than 0.35.

But if L is small enough then we cannot ignore ions falling on the end surface of the plates as well as difference of the plasma surface from cylindrical. Since the thinness of the plates is limited by structure durability the distance between the plates is limited as well. Thus this system will work well at low pressure when the width of the cathode layer is high enough.

2.2. Energy loss in neutralization

The energy of the neutrals will be determined by the difference of the potential energy of the particles between points of birth and neutralization. Ideally, if there was no resonant recharge and the reflection of the ions from the target surface would occur without loss, the beam of fast neutrals would be monoenergetic with the energy of particles determined by cathode voltage drop.

But the energy spectrum of the beam is influenced by two effects. Firstly, there is some loss of the energy in ion reflection from the target surface and this loss depends on of the incident angle of the ion to the target. Secondly due to the presence of the neutral molecules in the cathode potential drop region some of the ions may recharge on the gas target before reaching the cathode. Thus, besides the primary ions emitted from the plasma surface there will be some secondary ions formed by resonant recharge. As well as primary, secondary ions in the electric field of cathode potential drop will move to the cathode but their energy will be less. Moreover during the resonant recharge the neutrals are formed with energy determined by the potential difference between the plasma and point of recharge that is less than the energy of the neutrals formed by neutralization of the primary ions at the cathode.

Due to these processes, the energy spectrum of the neutrals beam spreads into the region of the low energies. This distortion depends on the neutral gas concentration in the cathode voltage drop region, the length between the plasma and the cathode and the energy loss by reflection from the cathode surface.

For the mathematical description of the process of ion neutralization by reflection from the solid target surface there are models of single and double binary collision [2]. Using the model of single binary collision the expression $E_i / E_n = (1+\mu^{-1})^{-2}(\cos(2\theta) - \sqrt{\mu^{-2} - \sin^2(2\theta)})^2$ for loss of energy was obtained, where μ - ration of the masses of the incident particle and atom of the target and θ - the angle of incidence related to the normal to the surface. Figure 4 shows graphs of energy loss on incident angle on the surface for the argon ions and few target materials. According to the obtained results the loss of energy increase with increasing angle of incidence of the ion to the target surface and the rate of the increase of the loss is greater the smaller the atomic mass of the target material. These suggest that the surface of the cathode-target should be made of a conductive material with high atomic number and the design of the cathode should be such that the ion incident angles to the cathode surface are small as possible.

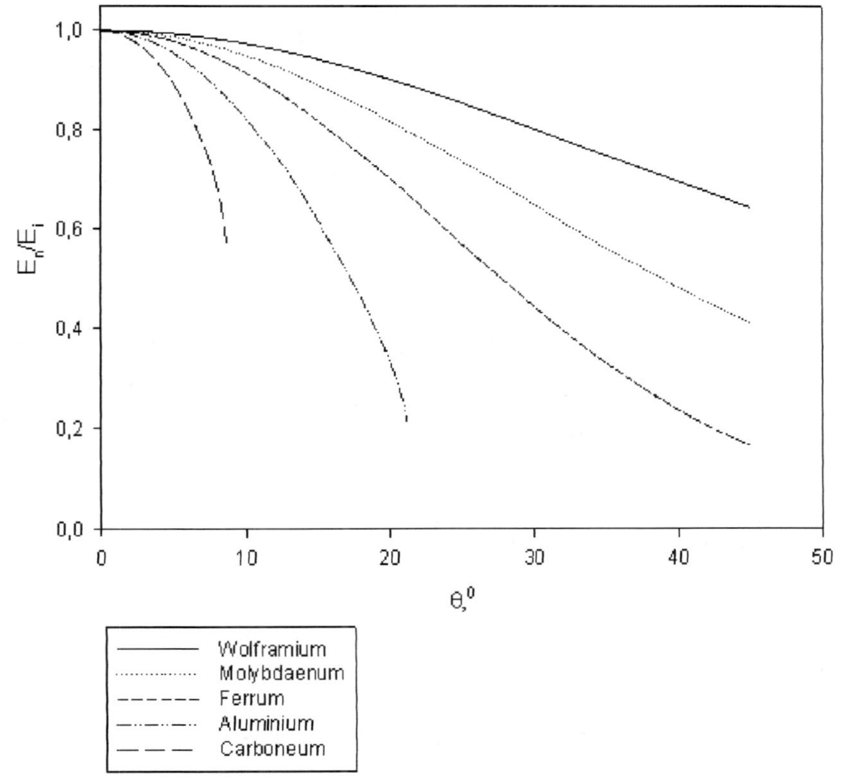

Figure 1. Energy loss in reflection

References

[1] Djagterev A V, Kudrja V P and Maishev J P 2009 *Microelectronics* **38** 188

[2] Koppers W R, Tsumori K, Beijersbergen J H M, Weeding T L, Kistemaker P G and Kleyn A W 1998 *Int. J. Mass Spectrum. Ion Physics.* **174.** 11

Journal of Physics: Conference Series **479** (2013) 012009 doi:10.1088/1742-6596/479/1/012009

The influence of transverse acoustic oscillations on contraction of the glow discharge

S A Fadeev and N F Kashapov
Kazan (Volga Region) Federal University, 18 Kremljovskaja str., Kazan 420008, Russian Federation

E-mail: fadeev.sergei@mail.ru

Abstract. The results of investigations of the interaction of longitudinal acoustic waves with electric gas discharge. Describes the phenomenon the formation of vortex flows in the acoustic gas discharge under the influence of a standing acoustic longitudinal waves. The phenomenon of occurrence vortex in the presence of transverse acoustic oscillations of the first tangential mode are analyzed. Presents the rationale of effective influence of the transverse acoustic oscillations on the glow discharge, which leads to the preservation of its diffuse form and, consequently, improve the energy input in the CO_2-laser.

1. Introduction

Laser – one of the most important inventions of the twentieth century. The most powerful lasers, continuous action, at the beginning of the twenty-first century, are CO_2-lasers. In addition to high output power CO_2-lasers have great efficiency up to 20%, and can effectively work in a pulsed mode. These features are responsible for CO_2-lasers diversity of applications: physical research, technological processes, location and communications.

Increased power deposited in the active medium of a CO_2-laser leads to the pinching of the positive column, the contraction of electrical gas discharge. When the contraction of the discharge spatial shape of the electron density distribution along the radius of the tube on the modified parabolic bell [1], there is a strong radial gradient of the gas temperature and the current density on the axis of the positive column gas temperature increases. Such discharge parameters do not meet the necessary conditions for work carbon dioxide laser with high specific power generation and lead to a complete cessation of laser generation. Discontraction discharge accompanied by the opposite phenomenon. Overcoming this kind of instability of the discharge is the main and most difficult problem in creating high-power lasers.

Conducted intensive research of the discharge in a longitudinal gas flow in order to overcome contraction of the discharge gap by increasing the gas pressure [2], to ascertain the possibility of withdrawal of heat from the tube, electrical shock emitted in the discharge area, and create a uniform uncontracted positive column at high pressures.

2. The influence of longitudinal acoustic oscillations on contraction of the glow discharge

Get discontraction discharge uniform at high pressures succeeded with the help of intense sound wave directed along the positive column. Experiments were carried out at the first resonant frequency of longitudinal acoustic oscillations [3]. When implementing the acoustic oscillations in the discharge gap CO_2-laser heat can be intensified, thereby reducing output time of heat from the discharge, which would raise the power density, and hence the density of the energy output with help radiation.

From the theoretical analysis of the processes that caused discontraction gas discharge sound wave follows that acoustic field inhomogeneity in the discharge tube there are, firstly, due to the boundary layer near the wall where the velocity is reduced from the value in the acoustic wave in the middle of the tube to zero at the wall of the tube, and second, due to the strong temperature gradient along the radius of the tube.

In the presence of solid walls, the acoustic field non-uniformity and the existence of the transverse profile of a standing wave arises steady vortex flow called acoustic flow [4], which reinforces transport processes along the radius of the tube. This causes an increase in the effective thermal conductivity of the plasma, leading to a decrease gas temperature increase of the diffusion losses of the charged particles on the wall and discontracted positive column.

For the existence of the phenomenon of acoustic flow is necessary that the size of the heterogeneity of the acoustic field was much larger than the thickness of the boundary layer. Thus, only the strong temperature non-uniformity across the radius of the tube is responsible for the occurrence of acoustic flows in a longitudinal standing wave affecting the transport processes [3].

From the foregoing follows that the discontraction discharge leads to equalization temperature gas on radius of the tube and, consequently, to reduce the degree of influence of the temperature non-uniformity on the acoustic excitation of the vortices.

3. The influence of transverse acoustic oscillations on contraction of the glow discharge

Longitudinal standing wave of acoustic oscillation is characterized by a specific location alternating maxima (anti-nodes) and minima (nodes) of the amplitude, while in the case of transverse acoustic oscillations set pattern is the same for each section a discharge chamber, where discharge light, thereby creating a uniform distribution of excited gas along the length of the tube. Research conducted on experimental stand [5] confirm the findings.

Experimental stand consisted of a molybdenum tube with an internal diameter of 9.5 cm and length 21 cm, closed at both ends faces of various shapes (conical, and hopping cross section) Electrode annular shaped with cross-sectional diameter of 0.4 cm and a DC power source. For creating sound waves used electrodynamic transducer, which was placed on the cylindrical wall of the end-hopping cross-section. In electrodynamic transducers was served a sinusoidal electrical signal with help generator. For the control parameters of the sound waves using a microphone which is able to move freely throughout inner surface of the tube, the signal from the microphone is fed to an oscilloscope. Reduced pressure created by the vacuum pump.

At realization the transverse acoustic oscillations in the tube should be expected discontraction at higher pressures and powers invested in discharge than in the case of longitudinal acoustic oscillations. Since by past the mechanism excitation of acoustic vortices, due to, the strong temperature non-uniformity, will be implemented transverse profile of a standing wave, which is an additional reason for the formation of acoustic vortices which carry the plasma to the radial direction and ensure the intensification of heat transfer along the radius of the tube.

Thus the velocity potential for transverse modes of acoustic oscillations has the form

$$\varphi = J_m(k_{mn}r)\cos(k_{mn}t \pm m\theta)e^{\varepsilon t}, \tag{1}$$

where J_m – Bessel function of m-th order, k_{mn} – wave number, ε – oscillation increment. Then the expression for the velocity of the transverse acoustic oscillations is

$$\boldsymbol{u} = -\frac{1}{k_{mn}}\frac{dJ_m(k_{mn}r)}{dr}\sin(k_{mn}t \pm m\theta)e^{\varepsilon t}\boldsymbol{e}_r \pm \frac{m}{k_{mn}}\frac{J_m(k_{mn}r)}{r}\cos(k_{mn}t \pm m\theta)e^{\varepsilon t}\boldsymbol{e}_\theta, \tag{2}$$

where $\boldsymbol{e}_r$ and $\boldsymbol{e}_\theta$ – the unit vectors, respectively, radial and tangential directions.

This is a acoustic solution for the transverse modes in a closed cylinder, which is a conservative system analogous to the discharge tube. Here the upper sign takes into account the growth of the amplitude of the time, the lower sign – decrease. Nonlinear generation of vortex flows is not oscillatory mechanism is mechanism a nonlinear stabilization of amplitudes of unstable oscillation.

Using the acoustic solution (2) to calculate the the stationary flows in the finite-amplitude acoustic waves. The mass flow in the fixed-point number of a closed cylinder

$$\boldsymbol{\mu} = \rho \boldsymbol{v}, \tag{3}$$

wherein ρ and $\boldsymbol{v}$ density and gas velocity at a fixed point, respectively. Represent parameters ρ and $\boldsymbol{v}$ as

$$\rho = \rho_0 + \rho, \ \boldsymbol{v} = \boldsymbol{v}_0 + \boldsymbol{v}_1, \tag{4}$$

ρ_1, $\boldsymbol{v}_1$ – acoustic components of density and gas velocity, respectively. For the mass flow with oscillations have

$$\boldsymbol{\mu} = \rho_0 (1+\eta)(\boldsymbol{v}_0 + \boldsymbol{v}_1) = \rho_0 \boldsymbol{v}_0 + \rho_0 \eta \boldsymbol{v}_0 + \rho_0 \boldsymbol{v}_1 + \rho_0 \eta \boldsymbol{v}_1. \tag{5}$$

Here $\eta = \rho_1 / \rho_0$ – dimensionless density. The first term of the expansion defines the average mass flow, terms of first order give corrections to momentum flux due to wave motion and because of the density fluctuations. Not the oscillating part of the secondary flow is only the second order term $\rho_0 \eta \boldsymbol{v}_1$. After averaging over the period of this term acoustic (with $\varepsilon = 0$) oscillations for tangential flow have the following result

$$\eta \boldsymbol{u} = \pm \frac{m J_m^2 (k_{mn} r)}{2 k_{mn} r} \boldsymbol{e}_\theta \tag{6}$$

The upper sign corresponds to the tangential wave propagating in a clockwise direction, the lower sign – wave propagating counterclockwise. The distribution of secondary the mass flow to the radius of the first tangential mode has a maximum radius at a distance of a quarter of the wall [6]. This result means that the acoustic waves create a steady transfer tangential pulse relative to the cylinder axis and as a consequence, gas transfer occurs.

4. Conclusions

That is, discontraction effect of the gas discharge sound wave can be enhanced by initiating acoustic oscillations with transverse velocity profile in the standing wave, thereby increasing intensification of heat transfer in the radial direction, and consequently, increasing the upper limit of energy input to the discharge, while maintaining them diffuse form that will to increase the power of the gas laser. This can be achieved by exciting in a discharge chamber transverse acoustic oscillations.

References

[1] Raizer Y P 1992 *Gas Discharge Physics* (Moscow: “Science”) p 536
[2] Kashapov N F and Israfilov Z K 1991 *JEPTER* **60** 364–367
[3] Aramyan A R and Galechyan G A 2007 UFN **177** 1207–30

[4] Landau L D and Lifshits E M 1988 *Hydrodynamics* vol 6, ed Y G Rudoi (Moscow: "Science") p 736
[5] Fadeev S A 2013 *Rep. sci. and tech. conf. "Low-temperature plasma in the process of application functional coatings"* (Kazan: Kaz. st. tech. un. press)
[6] Artamonov K I 1982 *Thermohydroacoustic stability* (Moscow: "Mechanical engineering") p 261

Protective matching polymer powder coating of piezoelectric element

V A Gavrilova, M G Fazlyyyakhmatov, N F Kashapov
Kazan (Volga Region) Federal University, 18 Kremljovskaja str., Kazan 420008, Russian Federation

E-mail: mfazlyjy@kpfu.ru

Abstract. Objects of research are coatings and technology of their applying to the piezoelectric elements for ultrasound. Acoustic impedance and thicknesses of matching layers for medical ultrasound transducers have been defined. In this paper performance characteristics of coating systems with predetermined properties have been selected. The conditions for selection of polymer powder paint for quarter wave matching layer have been determined. Conditions of forming polymer powder coatings have been proposed.

1. Introduction

Is known to in liquids and gases can only be distributed longitudinal elastic (acoustic) oscillations. This is due to lack of liquids and gases forms, and hence elastic form. Therefore, liquids and gases expand or contract only in the direction of wave propagation. Space filled with a substance in which the sound wave are spreading is called the sound field. The most important quantities characterizing the acoustic field are sound pressure (p) and the particle velocity (v) [1].

2. Calculation the thickness and the acoustic impedance of matching layer

The oscillatory particle velocity in the propagation of longitudinal acoustic waves is described by equation d'Alembert:

$$\nabla^2 \vec{v} - \frac{1}{V^2}\frac{\partial^2 \vec{v}}{\partial t^2} = 0. \tag{1}$$

Similarly, describes changing the acoustic pressure:

$$\nabla^2 \vec{p} - \frac{1}{V^2}\frac{\partial^2 \vec{p}}{\partial t^2} = 0. \tag{2}$$

Coefficient in front of second derivative in both equations is the inverse of the square of the velocity of propagation of longitudinal waves.

The joint solution of the equations defines the relationship between the amplitudes of the acoustic pressure (p_a) and particle velocity (v_a) of the longitudinal wave in the form of:

$$p_a = \frac{\rho_0 v_a \omega}{k} = (\rho_0 V) v_a, \tag{3}$$

where, ρ_0 - the equilibrium density of the medium, V - velocity of longitudinal waves.

The proportionality coefficient in brackets is called the acoustical impedance of the medium:

$$Z = \rho_0 V. \tag{4}$$

This quantity characterizes the environment in which the wave propagates, namely the scattering of wave energy in the sound field.

Ultrasound diagnosis, including medical, based on the phenomenon of reflection of elastic waves at the boundary between media with different acoustic impedances. However, the same phenomenon is one of the major challenges in the study since the acoustic impedance the source of ultrasonic waves and the object of the study may be different quite significantly, which causes strong reflection of ultrasonic waves while entering into the object. Presence of air between the source and the object under investigation is also fatal because of acoustic impedance values for metal and air are different an average of five orders of magnitude.

For successful research is needed to minimize losses when entering ultrasonic waves into the test object.

To eliminate of air are used couplant such as glycerin.

Reduction of losses upon reflection at the interface is the introduction of the quarter wavelength matching layer. This additional layer of material should have such acoustical impedance to provide equal acoustic impedance of the first medium and the second input acoustic impedance of the medium. Reflected from the both boundary of the matching layer waves will be equal in amplitude and antiphase that will lead to their mutual compensation. The phase shift of 180 degrees is provided the half wavelength path difference of the waves between the boundaries. Equality of the amplitudes - a rational choice acoustic impedance transforming layer [2].

The expression for calculating the acoustic impedance matching layer:

$$Z_{tr} = \sqrt{Z_1 Z_2}. \tag{5}$$

Based on the impedance value (5) material for the manufacture of the matching layer is chosen. Since the thickness of the matching layer is equal to a quarter wavelength in the material, the frequency band for which there will be a low reflection coefficient is small. To increase this frequency band is applied multiple matching layers.

Matching layers can be glued or can be electroplated. For the manufacture of the protective layer are selected materials characterized by high wear resistance, low attenuation of ultrasonic oscillations and high speed of sound, usually a sheet of a copolymer of vinyl chloride and ethyl acrylate [3]. Also, is used of industrial layers of a mixture epoxy resin filled with titanium dioxide. In the completed device case transition layer are made of the same material with the body. Transition layer is the integral part of the body.

The most common medical diagnostic ultrasound transducers are operating frequencies of 2, 2.5, 3.5, 5 and 10 MHz. Average propagation velocity of ultrasonic waves in human tissues is 1540 m/s, in water (at t = 20° C) 1480 m/s, in transformer oil 1390 m/s. Transformer oil is used to conduct simulation experiments, as the premises of piezoelectric elements in the water is not recommended by the manufacturer. The following table shows the thickness of the quarter wavelength matching layers for various piezoelectric elements.

Table 1. Thickness of matching layers piezoelectric elements for different media.

The operating frequency of the piezoelement, MHz	The thickness of the matching layer the piezoelement for humans, um	The thickness of the matching layer the piezoelement for water, um	The thickness of the matching layer the piezoelement for transformer oil, um
2	192.5	185	174
2.5	154	148	139
3.5	110	106	99
5	77	74	69.5
10	38.5	37	35

Selection of the coating is carried out on the basis of condition (5). For water, the acoustic impedance $Z = 1,483 \times 10^6\ kg/m^2\,\text{s}$, for piezoelectric ceramics PZT-19 $Z = 22,42 \times 10^6\ kg/m^2\,\text{s}$, respectively, the acoustic impedance the polymer powder coating should be about $Z_{tr} = 5,77 \times 10^6\ kg/m^2\,\text{s}$.

Acoustic impedance is defined by the expression (4). The velocity of propagation of elastic waves in the polymer powder coating is determined empirically.

3. The method of recovery the protective layer piezoelectric element

In operation, face of the transducer progressively wears away. In defectoscopy this problem is solved by replacing the prism normally set to the transducer. In medical diagnosis as a result of the properties of living tissues that are similar in properties to the water, prisms do not apply. In this case, the average transducer lifetime period is 2 years [4].

In [4] proposes the extension of lifetime the ultrasound transducer by applying a polymer powder coating on the surface of the piezoelectric transducer in the corona discharge.

The application of polymer powder coatings include coating and subsequent polymerization.

The spraying process is carried out by means of a spray gun – the unit for the application of polymer powder coating. To the electrodes spray is applied a high voltage. And between the spray gun and the grounded detail is created a strong electric field. Powder particles entering the region of corona discharge, acquire a charge [5].

The spraying process takes place in the spray chamber, designed to prevent scattering and centralized collection powder coating particles are not deposited on the painted product.

Experiments have been conducted and optimum operating parameters is determined of the system of homogeneous coating polymer powder coatings (PPC) (Table 2).

Table 2. Optimal operating parameters of the PPC application.

Thickness of the matching layer, um	Spraying conditions			
	The voltage on the discharge electrode, kV.	Supply air pressure, atm.	The distance from the needle spray gun, cm	Time of the coating process, s
192.5	25	0.8	15	8
154	25	0.8	30	6
	30	0.8	25	6
	30	0.6	20	6
110	10	0.8	30	6
77	10	0.4	25	2
38.5	20	0.4	15	2

The polymerization process occurs in the drying chamber, where due to the high temperatures, is the heating of products, fusion of the polymer particles and the final formation of the coating. [6]

Particularly important adherence solidification thermosetting paints, because any deviation from adversely affecting the properties of the resulting coatings.

For medical products is used polyether, epoxy- polyether coating. Their main characteristics are: abrasion resistance, impact resistance, surface smoothness and comfort, the lack of toxicity, obstruction of bactericidal activity, suitability for sterilization, easy to clean, resistant to chemical attack [7].

Table 3 shows the conditions of forming polymer powder coatings.

Table 3. The conditions of forming PPC.

Type of PPC	Temperature of curing, ° C	Time of curing, min
	180	15
Polyether	200	10
Epoxy-	180	15
polyether	200	8

In Figure 1a. is presented transducer with the removed protective cover. The outside of the piezoceramic plates electroplated nickel-plated. The result of applying a protective coating is shown in Figure 1b.

Figure 1a. Transducer with the removed protective cover.

Figure 1b. The result of applying a protective coating.

4. Conclusions

Performance characteristics of the coating system with predetermined properties have been selected. The conditions for selection of polymer powder coating have been determined. The conditions for the formation of polymer powder coatings of various types have been proposed.

References

[1] Golyamina I P 1979 *Ultrasound. Little encyclopedia* (Moscow: "Soviet encyclopedia") p 400
[2] Solovjaninova I P and Shabunin S N 2004 *Theory of wave processes* (Yekaterinburg: "GOU VPO UGTU-UPI") p 142
[3] Zacepin A F 2006 *Acoustic control* (Yekaterinburg: "GOU VPO UGTU-UPI") p 117
[4] Gavrilova V A, Kashapov N F, Kashapov R N 2011 Biomedical Engineering **45** 198-200
[5] Alekseev V A, Israfilov Z K, Kashapov N F, Razzhivina L G, Solov'eva I N 2000 A Device for Dispersed Material Coating RF Patent No. 2146563
[6] Gavrilova V A, Kashapov N F 2010 Vestnik KGTU **7** 117-125
[7] Yakovlev A D 1987 *Powder paints* (Leningrad: "Chemistry") p 216

Research of the impact acidity of electrolytic cathode on the course of the plasma-electrolytic process

L N Kashapov[1], N F Kashapov[1] and R N Kashapov[1,2]
[1] Institute of Physics, Kazan Federal University, Kazan, 420008, Russia
[2] Laboratory of Radiation Physics, Kazan Physical-Technical Institute of the Kazan Scientific Center of the Russian Academy of Sciences, Kazan, 420029, Russia

E-mail: kashramiln@gmail.com

Abstract. The object of research is a gas discharge with an electrolytic cathode, namely the influence of the nature of the cathode on the discharge. Is established that with decreasing the acidity of electrolytic cathode voltage of discharge ignition decreases. A mechanism is proposed to maintain the discharge at the gas-electrolyte interface.

1. Introduction

The classic methods of electrical treatment of steels, such as electrochemical, EDM, electro-thermal and electromechanical have drawbacks. Namely, a high consumption of electrical energy, big problems of disposing of the reaction products, the difficulty of obtaining the required surface profile at a given scale. In this regard, a strong demand existed develop and research new environmentally-friendly and cost-effective methods of electrical processing.

One of the new electro physical types of processing is the plasma-electrolytic method [1]. This method of processing is carried out by applying a voltage to electrodes in an electrolytic solution as a result of around one of the electrodes is initiated gas-vapor discharge [2]. The peculiarity of this discharge is that it burns between a solid and a liquid. Solid electrode can be both the anode and cathode. When discharge burning occurs at the cathode cleaning and polishing a workpiece, also under certain circumstances may be application of coatings [3]. When the anodic process on a metal having the property of forming during anodic polarization protective oxide layer with a high resistance is anodized with microarcs combustion, the process is called oxidation microarc [4]. On metals not having given properties observed erosion destruction of the metal and dissolution. The most similar in their impact on the work surface is a combined method of electroerosive and electrochemical machining [5]. In which the metal removal is occurs due to the anodic of dissolution and the electrical erosion. Anodic dissolution is the electrochemical oxidation of the metal in the electrolyte solution to form a soluble products and discharge of the liberated electrons to the external circuit. The difference between this combined method of processing from the anodic process of plasma-electrolytic processing is that erosion damage occurs due to the breakdown of the electrolyte and the formation of a discharge channel between the workpiece electrode (anode) and the tool electrode (cathode), and not due to burning gas discharge between the metal anode and electrolytic cathode. This is due to the fact that in the plasma-electrolytic processing interelectrode space is increased from 0.5 mm to tens of centimeters and is applied to the electrodes of more high voltage value (U>30 B). Despite such a broad application of the variety of discharge, remains an open question about the mechanism of maintenance of the discharge. The answer to this question is related with the mechanism of electron emission from the cathode electrolyte - a liquid. The emission of electrons, of course, will depend on the type of electrolyte and the aim of this work was an experimental research of influence of acidity (hydrogen index) on the course of plasma-electrolytic process.

Of interest is the effect of the parameters of the solution to the ignition of the gas discharge with an electrolytic cathode. One of the important parameters of the electrolyte is its acidity, namely hydrogen index, pH - a measure of the activity of hydrogen ions in solution, and quantitatively expresses its acidity is calculated as the negative (taken with the opposite sign) decimal logarithm activity of hydrogen ions, expressed in moles per liter: $pH = -\lg(H^+)$. Some minimal spread received reverse pH value - index of basicity of the solution, pOH, equal to the negative decimal logarithm of the concentration in the solution of ions OH^-: $pOH = -\lg(OH^-)$ as in any aqueous solution at 25 °C $[H^+][OH^-] = 1{,}0\times10^{-14}$, it is clear that at this temperature: $pOH = 14 - pH$.

2. Experimental

To achieve this goal has been applied experimental setup which described in [5]. Was used pulsating current source obtained after full-wave rectification with smoothing capacitor filter and without it. Changing the shape of current and voltage at the time of discharge ignition was determined by the oscilloscope FLUKE 105 SCOPEMETER SERIES II, timebase varied from 5 ns to 60 s. The relative measurement error did not exceed 0.025%. Measurement of voltage and current discharge was carried out using two digital universal measuring devices MMH-930 and APPA 109 N relative measurement error is 0.8%. The measurement system was automated and included the analog input module measuring MBA8 and Digital Multimeter APPA 305 connected to the computer. To prevent changes in the size of the area of the anode wore dielectric protective case which leaves open only a well-defined part of the electrode. The electrodes used graphite anode and a copper cathode. S_{anode}=316,25 mm^2, $S_{cathode}$=1802 mm^2, ($S_{anode}/S_{cathode}$ =0,175). In this case, the discharge burns at the anode, which corresponds to the statement [5], where noted that plasma-electrolytic process takes place when $S_K > 5 S_A$.

3. Results

For of experimental research by using smoothing capacitor filter were prepared 4 electrolytes: 5% aqueous solution of acid H_2SO_4 pH = 2; 5% aqueous salt solution Na_2SO_4 + 2% H_2SO_4 pH = 4; 5% aqueous Na_2CO_3 pH = 7; 5% aqueous alkaline solution NaOH pH = 11. The acidity of the solution changed in two ways: choice of type of electrolyte and changing the concentration of the electrolyte solution. Was investigated the dependence of the value of the ignition voltage of the discharge from the acidity of the solution. It was found that with increasing pH and decreasing ignition voltage of the discharge. This dependence is shown in Figure 1.

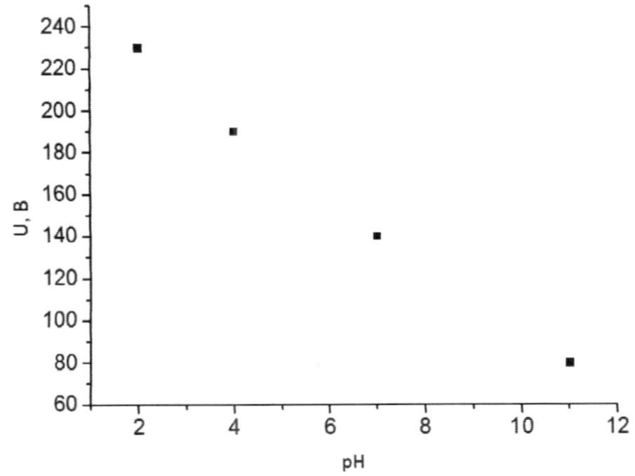

Figure1. The dependence of the ignition voltage of the discharge from pH.

As can be seen increasing the concentration of hydroxide-ions OH^- reduces the ignition voltage. This suggests that the necessary amount the charge carriers (electrons) is reached at lower voltages for solutions with a higher concentration of OH^-. Thus it is possible to speak of hydroxide-ion plays an important role in maintaining the discharge. It was also revealed decrease pH with time discharge burning, which is observed in [6].

Experiments were conducted to determine the influence pH of solution (electrolytic cathode) on the current-voltage characteristics (CVC) of plasma-electrolytic process. In these experiments, we used the power supply without the smoothing capacitor filter. During the experiments, was fixed at which voltage rise to strong sound waves and discharge initiation.

Figure 2 shows the current-voltage characteristics of plasma-electrolytic process for various electrolytes. Current-voltage characteristics increase smoothly at the beginning and peaking are down. On a site where there is a linear increase in current on the voltage dominates the electrochemical processing. In the area of 65 - 120 V (depending on the type of electrolyte) begin to occur anode microdischarges and observed the formation of gas bubbles. With decreasing the current is observed increase in the intensity of the glow of plasma-electrolyte discharge for the salt solutions and bases. Established that the addition of low concentrations of alkaline or acid electrolyte hasn't a strong influence on the form the CVC, but adding more alkaline electrolyte the CVC maximum is shifted to lower voltages and drops from 112V to 70V.

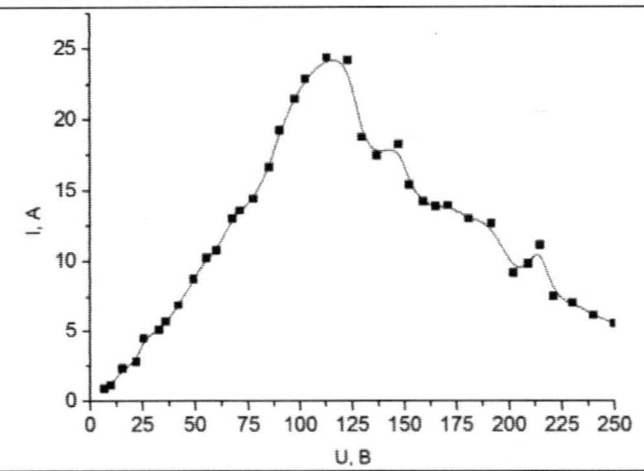

Figure 2a. The CVC plasma electrolytic process in solutions 5% Na_2SO_4 (200 ml.)

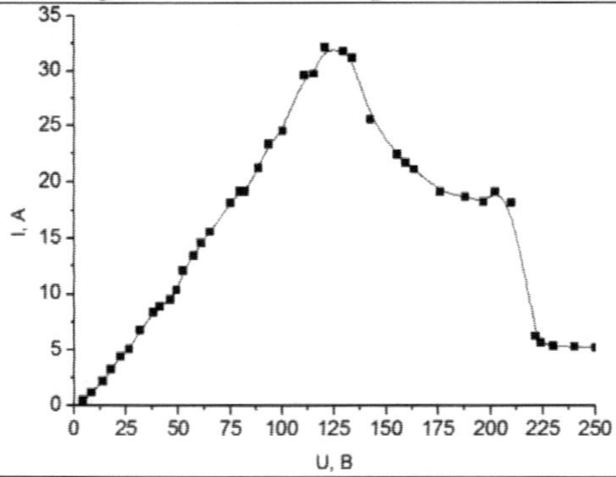

Figure 2b. The CVC plasma electrolytic process in solutions 5% Na_2SO_4 (200 ml.)+ 5% NaOH (2 ml.)

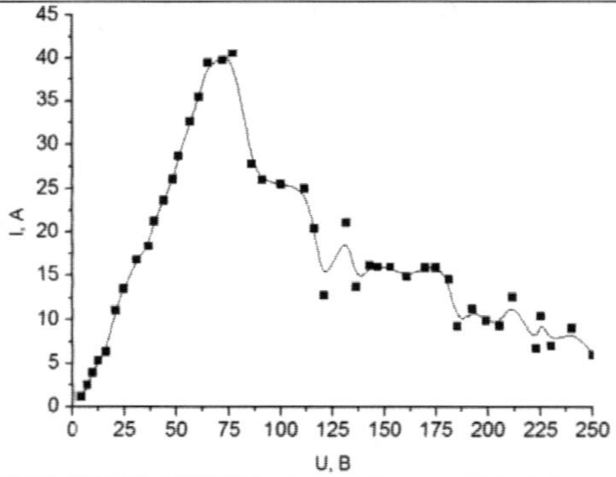

Figure 2c. The CVC plasma electrolytic process in solutions 5% Na_2SO_4 (100 ml.)+ 5% NaOH (100 ml.).

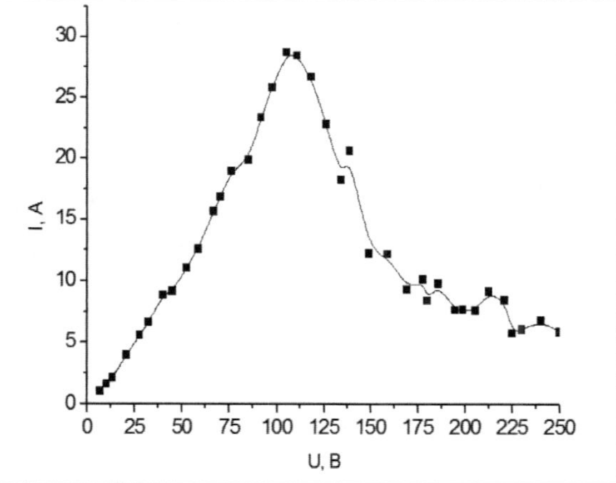

Figure 2d. The CVC plasma electrolytic process in solutions 5% Na_2SO_4 (200 ml.)+ 2% H_2SO_4 (2 ml.).

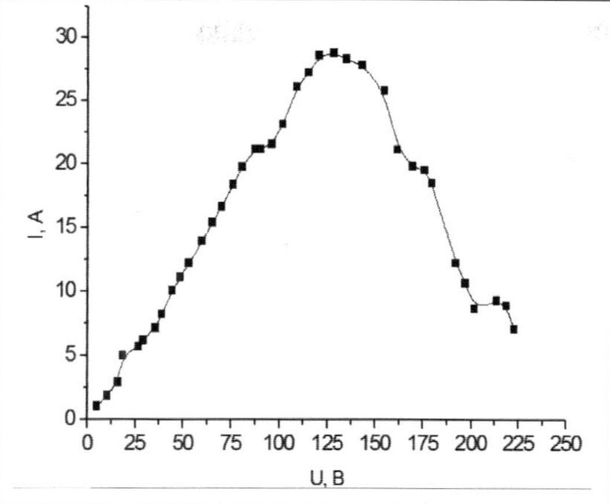

Figure 2e. The CVC plasma electrolytic process in solutions 5% Na_2SO_4 (100 ml.)+ 2% H_2SO_4 (100 ml.).

In the case of adding an acidic electrolyte opposite picture is observed and the maximum the CVC shifting from 112 to 130 V. The first CVC describe the plasma-electrolytic process in a neutral environment. Found that sound waves observed at ~ 75 V and glow occurs ~ 90-95 V. For environment 5% Na_2SO_4 (200 ml.) + 5% NaOH (2 ml.) with indicator pH = 10 as sound waves observed at ~ 75 V, glow ~ 90-95 V. But for the 5% Na_2SO_4 (100 ml.) + 5% NaOH (100 ml.) with indicator pH = 12 with sound vibrations ~ 41 V, glow ~ 65 V. For 5% Na_2SO_4 (200 ml.) + 2% H_2SO_4 (2 ml.) and 5% Na_2SO_4 (100 ml.) + 2% H_2SO_4 (100 ml.) sound vibrations at ~ 85 V, glow ~ 120 V. Thus, with the pH change of the solution occurs CVC maximum displacement.

As can be seen, hydroxide- ion plays an important role in maintaining the plasma discharge. And it can be assumed that the discharge will be to certain the limiting reduce of concentration OH^-. Also in [7] was founded that when the discharge was is observed super Faraday formation of hydrogen peroxide H_2O_2. Hence we can assume the following mechanism for the maintenance discharge at the gas-electrolyte:

$$OH^- - e \rightarrow OH$$
$$2\,OH \rightarrow H_2O_2$$

Acknowledgments
The authors gratefully acknowledge the financial support from the Russian Foundation for Basic Research №13-08-00906 A.

References

[1] Slovetskii D.I., Terent'ev S.D., Plekhanov V.G. 1986 Thermal physics of the High Temperature **24** 353-359..
[2] Slovetskii D.I., Terent'ev S.D. 2003 High Energy Chemistry **37** 355.
[3] Meletis E.L., Nie X., Wang F.L., Jiang J.C. 2002 Surface and Coating Technology **150** 246.
[4] Suminov I.V., Epelfeld A.V., Lyudin V.B., Crete B.L., Borisov A.M. Microarc oxidation (the theory, technology, equipment Textbook. 2005 368
[5] Kashapov R.N. 2010 Physics and Chemistry of Materials Processing **5** 50-56.
[6] Khlyustova A.V., Maximov A.I., Titov V.A. 2004 High Energy Chemistry **3** 227 – 230
[7] Polyakov O.V., Badaljan A.M., Bakhturova L.F. 2003 High Energy Chemistry **5** 367-372.

Dust particles of plasma of the vacuum arch - structure and properties

N A Smolanov and N A Pankin

Ogarev Mordovia State University

E-mail: smolanovna@yandex.ru

Abstract. As a part of plasma of the arc category there are electrons, ions and particles, including drop fraction. On these signs it can be carried to dust plasma. In work the analysis of okolokatodny processes is presented in the arc category in the presence of the magnetic and electric fields influencing reversible nature of movement of a cathodic spot and a plasma stream as a whole. Results of element and granulometric composition of the black powders received from plasma of the arc category at dispersion of the titan, their magnetic properties are given.

1. Introduction. It is known that the metal drops leaving the cathode, intensively evaporate and turn into plasma educations already in close proximity to a cathode [1] surface. At drop movement in a plasma stream in it there can be a convective mass transfer, the drop is under pressure streams of ions and electrons, and in certain cases there is a rotation of drops and their destruction (explosion).

The processes proceeding in a cathodic spot and dust plasma at imposing of an external magnetic field, have identical character. The particular interest represents retrograde behavior of a cathodic spot when at a certain value of an external magnetic field the cathodic spot of a vacuum arch moves in opposite to Ampere force the direction [1-4]. Let's note that change of the direction of rotation of a cathodic spot – known long ago, but the phenomenon not learned up to the end.

Interest to behavior of a cathodic spot in externally magnetic field is connected with the similar behavior of dust plasma found in a number of works in a magnetic field [5-6]. The existence reasons at a dust granule of the magnetic moment - element structure and generation of the magnetic moment by superficial loopback current of the rotating loaded particle. It is established that depending on the size of a magnetic induction magnetic properties of dust plasma are defined by various factors. In a weaker magnetic field замагничена only electronic a plasma component, in average fields (0.1–0.3 T) ions, magnetized and only in strong fields (2–3 T)_magnetized dust granules. In process of increase in an induction the stop of rotation of dust plasma with the subsequent change of its direction is possible. However the reason of threshold nature of emergence of rotation, as well as in case of the return movement of a cathodic spot, remains until the end of not clear [5-6].

2. Work purpose. Research of morphology, phase, element structure and distribution by the sizes melkodispersny the powder which has settled on walls of the vacuum camera at dispersion of the titanic cathode by an arc method. In work the element and granulometric composition of the black powders received from plasma of the arc category at dispersion of the titan in the atmosphere of reactionary gases was investigated. For the purpose of clarification of influence of magnetic and electric fields on structure and properties, melkodispersny powders received at various operating modes of cathodic knot – the potential of a substrate and current of the focusing magnetic coil changed.

3. Receiving and research technique. Dust-like particles (fig. 1,2) were besieged on walls of the vacuum camera of the NNV-6 installation in the course of ion-plasma drawing multilayered film coverings on substrates (steel 12X18H10T). Technological modes of sedimentation: arch current – 75 A, shift potential —250 V, or 0 V (the anode camera walls were). At a warming up and ionic cleaning – 500-600 V. Material of the cathode – a titanic alloy of VT1-00. As carboniferous components the acetylene given to area of the arc category of titanic plasma in the ratio with nitrogen about 4:1 was used. Pressure of a mix of nitrogen and acetylene – 0,1 – 0,05 Pas. Adjustment of supply of gases in the conditions of dynamic vacuum was made by RPG-1/1 sensors with feedback with the vacuum gage. It provided rather stable pressure (within 10-2 Pas + - 15%) in the camera during time (20 min.) all process.

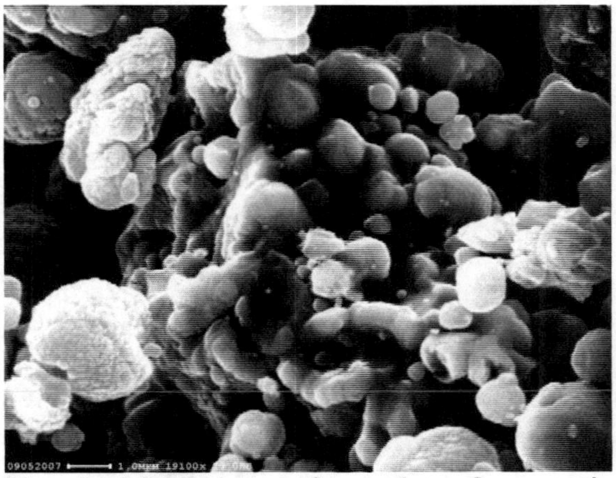

Fig. 1. Morphology of powder of magnetic fraction (x19 000)

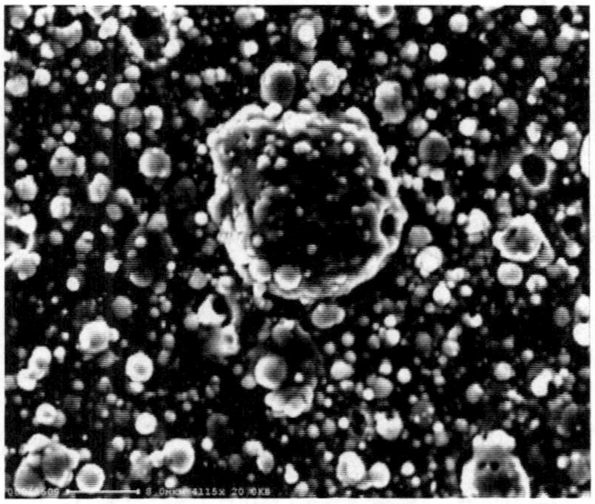

Fig. 2. Electronic microscopy of the condensed drop phase increase (x4115)

(Fraction less than 0.075 mm) extracted powder from walls of the vacuum camera the vacuum cleaner and subjected to magnetic separation in the field with intensity to 1000 E. Granulometriya of particles was made on the analyzer of the sizes of particles of Shimadzu SALD – 3101. Ranges of EPR were registered on a radio spectrometer of PS100X (frequency of 9.3 GHz). Microstructure and element structure investigated on a raster electronic microscope of LETI at an accelerating voltage of 20 kV.

4. Results. The dispersion of particles doesn't give the chance to determine by the sizes precisely magnetic properties, averaging them. It is known that for particles with small sizes jump of properties or their nonlinear dependence are characteristic at some size. The carried-out comparison of dispersion of the particles received at various operating modes of cathodic knot (interelectrode potential, current of the focusing coil) showed unequal nature of their distribution. For the particles collected from walls of the vacuum camera, distributions of type Gaussian, or described by decreasing sedate function are characteristic. The type of distribution depends, first of all from parameters of process of sedimentation. For example, the most characteristic type of distribution by the sizes of not magnetic particles for a mode with giving on a substrate of the potential of-250 V is shown in fig. 3. The sizes of the particles received at giving of potential on the submarine before magnetic separation, are in limits 5-100mkm. At sedimentation of particles without potential giving (the anode – walls of the vacuum camera) their main number is in limits of 70-130 microns.

EPR analysis - ranges showed (fig. 4) that powders are characterized by magnetic properties with various g-factors. Feature of a range of EPR of studied powders is its asymmetry. Let's note also that with reduction of the sizes of fraction in a range of absorption there are additional lines. Magnetic separation also leads to emergence of new lines in a range of absorption by particles of the magnetic and low-magnetic fractions which aren't observed in an initial condition.

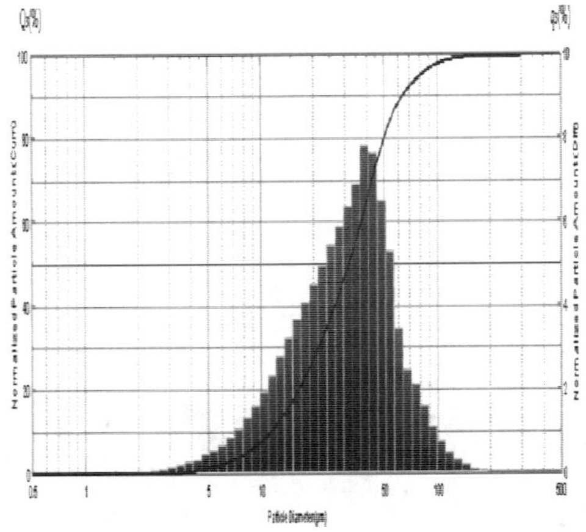

Fig. 3.raspredeleny by the sizes of black particles (U=250 V)

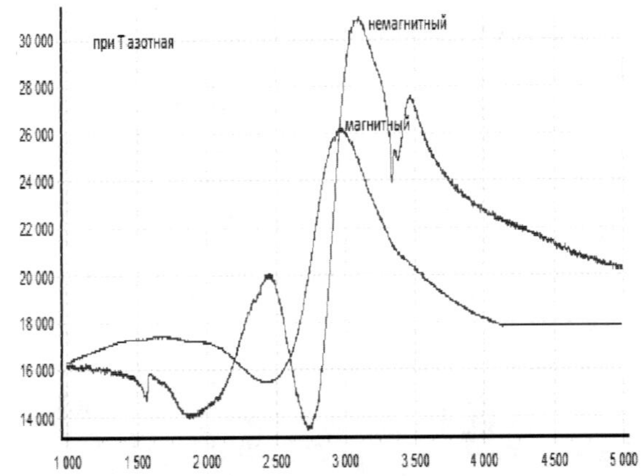

Fig. 4. EPR - ranges of powders at a nitric temperature

We considered possible mechanisms of evolution of dust plasma from a cathodic spot to its recombination on a substrate and walls of the vacuum camera in the conditions of simultaneous imposing of fields. Crystallization of a drop component of plasma and ionic streams on walls of the vacuum camera leads to education micro – and nanostructures, i.e. to manifestation of condensation of substance in wild spirits. Thus there can be structures of unknown modifications [7, 8]. In received structure from drops there are firm solutions and the "unusual" [8] crystal phases caused by nonequilibrium conditions of their receiving – with a fast speed of transition of couples liquid-crystal or steam - a crystal, with imposing E and N. Eto leads to nonequilibrium plasmochemical processes and formation of microstructures with a big free surface (see fig. 1,2).

5. Discussion and conclusions. On the basis of the conducted researches we drew a conclusion that studied melkodispersny powder can be considered as metallic structure, including in a look эндофуллеренов. The black structure can have variable element and multiphase structure (see tab. 1).

Tab. 1

Element composition of powders (with a potential of substrate of-250 V)

fraction	Element composition of powders, ат. %					
	Ti	N	O	C	Fe	Cr
not magnetic	44.23	9.69	35.29	10.30	0.12	0.03
The magnetic	51.60	7.40	29.97	10.21	0.26	0.10

In condensate the reactionary gases forming difficult, even unknown hydrocarbonic connections, including with the titan are occluded. Structure germs probably are microdrops of Ti of the various sizes, arising in a cathodic spot and crystallizing then on walls of the vacuum camera in nonequilibrium conditions. The structure of powder can contain also elements of the substrate sprayed in the course of ionic cleaning. Dust particles can be divided into 2 groups. The first group - (slow) - is kept by a gasdynamic stopper and settles near the cathode and on substrates in the camera. The second - fast ions and the small drops formed in a cathodic spot at issue, create from the raised condition of nanostructure at fast (an adiabatic mode) coolings on walls of the vacuum camera. The question of mechanisms of paramagnetism of these microparticles is discussed.

Journal of Physics: Conference Series **479** (2013) 012012 doi:10.1088/1742-6596/479/1/012012

References

[1] Lyubimov G A, Rakhovsky V I 1978 *Uspekhi Fizicheskikh Nauk*. **125** 665 (in Russian)
[2] Andreev A A 2007 *Izv. Kharkov. University*. **32** 763. (in Ukraina)
[3] Barengolts S A, Litvinov E A 1998 Technical Physics. **68** 60. (in Russian)
[4] Beavers Yu K, Bystrov V P, Rukhadze A A 2006 *Technical Physics* 76 35. (in Russian)
[5] Sextons L G 2011 FNTP-2011. **1** 140 (in Russian)
[6] Karasev V Yu, Dzliyeva E S 20011 FNTP-2011. **1** 147(in Russian)
[7] Pleskachevsky Yu M 2008 *Nanoparticles in the condensed environments* (Minsk: Prod. BGU center). (in Belorus)
[8] Andriyevsky R A 1997 *Successes of chemistry*. **66(1)** 57. (in Russian)

Journal of Physics: Conference Series **479** (2013) 012013 doi:10.1088/1742-6596/479/1/012013

Calculation of the composition of the equilibrium dusty plasma

I Fayrushin[1] and G Dautov[2]

1 Department of Technical Physics and Energetics. Kazan Federal University, Kazan, Russian Federation

2 Department of General Physics. Kazan National Research Technical University after name A.N.Tupolev – KAI, Kazan, Russian Federation

E-mail: fairushin_ilnaz@mail.ru

Abstract. We consider the equilibrium dusty plasma consisting of neutral atoms of the gas, the gas ions, free electrons and charged dust particles. We obtained the generalized Saha-equation for equilibrium dusty plasmas. In the approximation of a rectangular potential well at the surface of dust grains are calculated charge of the dust particles, the concentration of ions and electrons. It is shown that depending on the basic parameters of the plasma charged dust particles can be both positive and negative.

1. Introduction

Dust equilibrium plasma is formed in the process of combustion of various fuels, with functional coatings applied by thermal plasma flow and plasma chemical reactors of nano-and micro-sized particles [1-6]. Particles condensed substances called particulates or dust particles in the plasma at atmospheric pressure are in the statistical balance with their surrounding electrons, ions and neutral atoms. This mixture is generally electrically neutral system.

2. Theoretical background

We assume that the concentration of particulate matter is n_r, the average concentration of electrons in a gas or plasma is n_e, one has a particulate charge in units of electron charge is equal to N, the electron gas and particulates are in a state of statistical equilibrium. Also, we introduce the following notation: z_i and z_a - the partition on the electronic states of the ion and atom respectively, φ_u - single ionization potential of the gas atom, q - the absolute value of the electron charge, n_i - ion concentration, n_a - the concentration of atoms, h - Planck's constant, θ - statistical temperature is equal to kT, k - Boltzmann constant, T - absolute temperature, m_e - mass of the electron.

From the law of mass action:

$$\frac{n_i n_e}{n_a} = \frac{2z_i}{z_a}\left(\frac{2\pi m_e \theta}{h^2}\right)^{\frac{3}{2}} \cdot e^{-\frac{e_1\varphi_u}{\theta}}. \quad (1)$$

and balance of charged particles: $n_r N + n_i = n_e$ obtain

$$\frac{x_i(n_r N + x_i n)}{1-x_i^2} = \frac{2z_i}{z_a}\left(\frac{2\pi m_e \theta}{h^2}\right)^{\frac{3}{2}} \cdot e^{-\frac{e_1\varphi_u}{\theta}}, \tag{2}$$

where $x_i = \dfrac{n_i}{n_a + n_i}$ - the degree of ionization, $n = n_e + n_i + n_a$ – the total electron density of the plasma ions and atoms.

Thus, (2) is a generalization of the Saha-equation, taking into account the impact of particulates on the degree of ionization.

The condition of statistical equilibrium [7] of the electron gas in a dusty plasma is expressed by the following equation

$$\mu_1 + e\varphi_1 = \mu + e\varphi,$$

where μ_1 - to the Fermi energy of the electronic gas inside in dust particle and μ - the Fermi energy of the electronic gas outside of dust particle, φ_1 and φ - the potentials of the electric field inside and outside the particulates, e - electron charge. We assume that $\varphi = 0$ and inside dust particle $\varphi_1 = const$, $|e| = q$. With this in mind, equation (1) can be written as

$$\mu_1 - \mu = q\varphi_1. \tag{3}$$

from (12) we obtain

The potential energy of an electron inside the dust particle, the charge is equal to Nq, is determined by the known equation

$$W_n = -q\varphi_1 = -\frac{q^2 N}{4\pi\varepsilon\varepsilon_0 R}. \tag{4}$$

where ε - relative dielectric permittivity, ε_0 - electric constant.

The concentration of free electrons in a gas or plasma is usually low, and therefore as μ it necessary to use the Fermi energy of a nondegenerate electronic gas

$$\mu = \theta \ln(a n_e),\ a = \frac{1}{2}\left(\frac{h^2}{2\pi m_e \theta}\right)^{\frac{3}{2}}. \tag{5}$$

Depending on temperature and concentration of electrons in dusty particles are two possibilities. For large n_{e0} and small T the electronic gas in a dusty particle is degenerate, while for small n_{e0} and large T it is nondegenerated. Lets consider second case in more detail. The Fermi energy of the electronic gas in a dusty particle is defined by

$$\mu_1 = \theta \ln(a n_{e1}), \tag{6}$$

where n_{e1} - concentration of the electrons in a dusty particle after emission.
Substituting the expressions (4), (5) and (6) into (3) gives

$$\theta \ln\frac{n_{e1}}{n_e} = bN,\ b = \frac{q^2}{4\pi\varepsilon\varepsilon_0 R}.$$

$$\frac{n_r N + n_i}{n_{e0} - \dfrac{N}{V}} = e^{-\frac{bN}{\theta}} \tag{7}$$

3. Results

Using equations (2) and (7) can perform calculations of electrons and ions in plasma dust depending on the radius of the dust particles, their concentration and the temperature of the plasma. The results are presented in Fig. 1-6. It is assumed here that the total volume fraction of all dust particles is constant and amounts to 4% of the entire plasma, $n_{e0} = 10^{19}$ m^{-3}, the gas pressure is 10^5 Pa, $\varphi_u = 4{,}3$ eV, which corresponds to the argon.

As seen from these figures the electron density of the plasma at a temperature of 1000 K and 1500 K with increasing radius grains decreases. This phenomenon can be explained by the fact that the electron yield is strongly dependent on the radius of the dust particles, smaller particles efficiently emit and absorb electrons. Since the electrons emerging from the plasma with grains increasing *R* becomes smaller, as expected, an increase in ion concentration associated with a reduction of their recombination with electrons (Fig. 2 and 4).

Another character of the observed at 2000 K. This is due to the fact that at the higher temperature, the concentration of their own plasma electrons generated in the process of ionization of the gas is greater than the initial concentration of electrons in the dust particles . Dust particles are smaller with increasing radius capture electrons , so the concentration of ions in the plasma decreases (Fig. 5,6). At smaller values of the radius of dust more electrons from the plasma enters inside of the dust particles , thus the concentration of ions decreases.

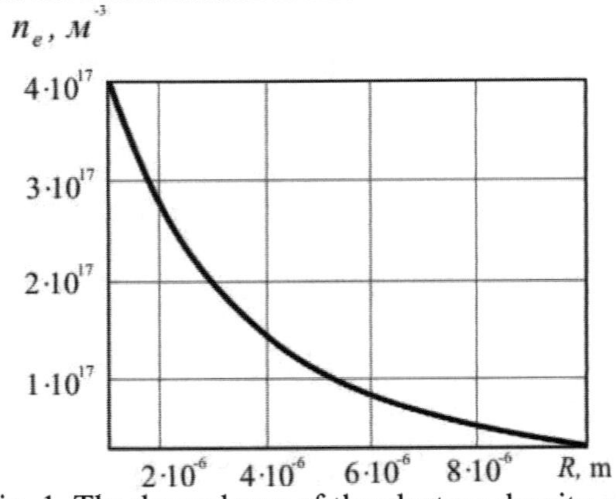

Fig. 1. The dependence of the electron density of the plasma radius of dust particle matter at T = 1000 K.

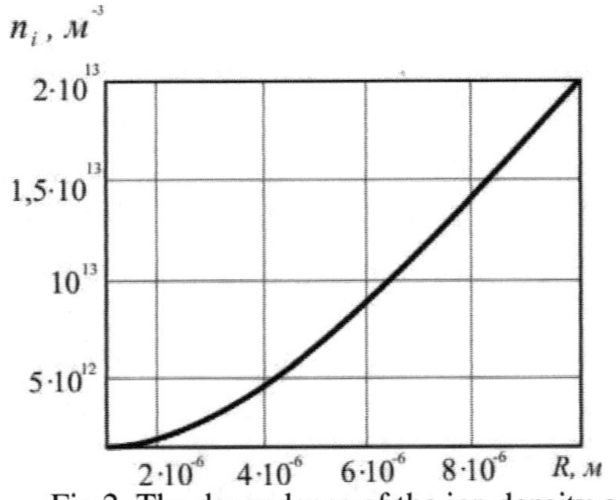

Fig.2. The dependence of the ion density of the plasma radius of dust particle matter at T = 2000 K.

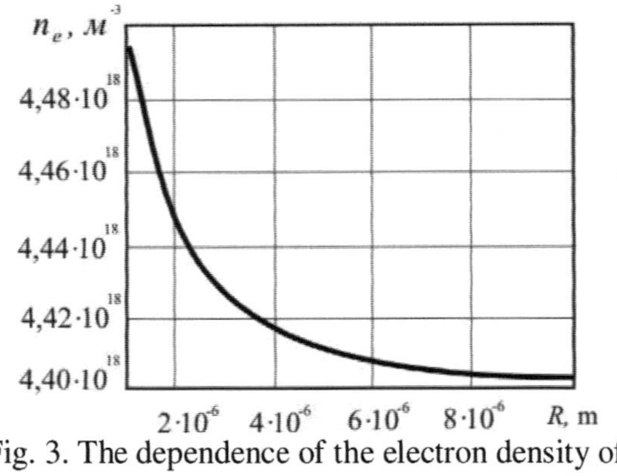

Fig. 3. The dependence of the electron density of the plasma radius of dust particle matter at T = 1500 K.

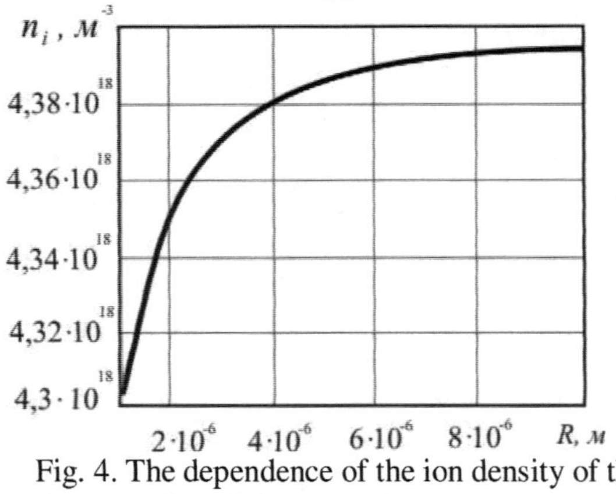

Fig. 4. The dependence of the ion density of the plasma radius of dust particle matter at T = 2000 K.

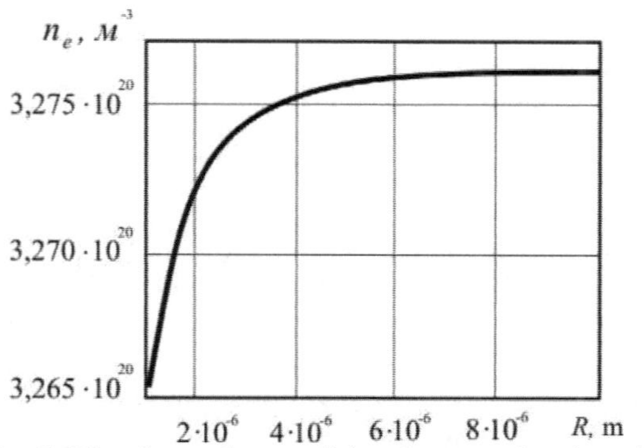

Fig. 5. The dependence of the electron density of the plasma radius of dust particle matter at T = 2000 K.

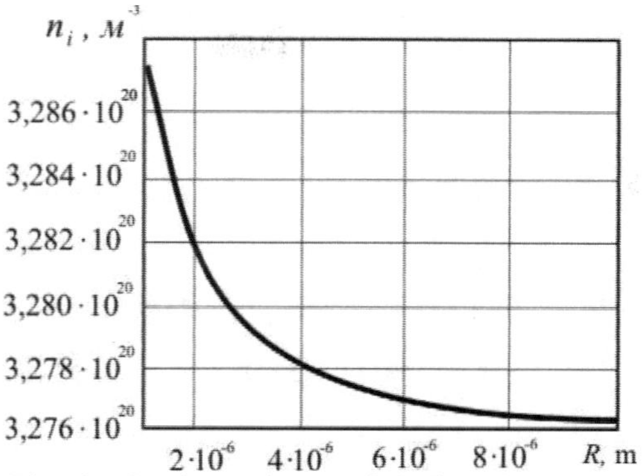

Fig. 6. The dependence of the ion density of the plasma radius of dust particle matter at T = 2000 K.

From Fig. 7 that the charge of the dust particles in absolute terms decreased linearly as a function of their radius.Thus, the specific charge of the particles with a decrease in R will increase, which confirms the above assumption about the growth of the efficiency of the emission or absorption of electrons with a decrease in the size of dust particles.

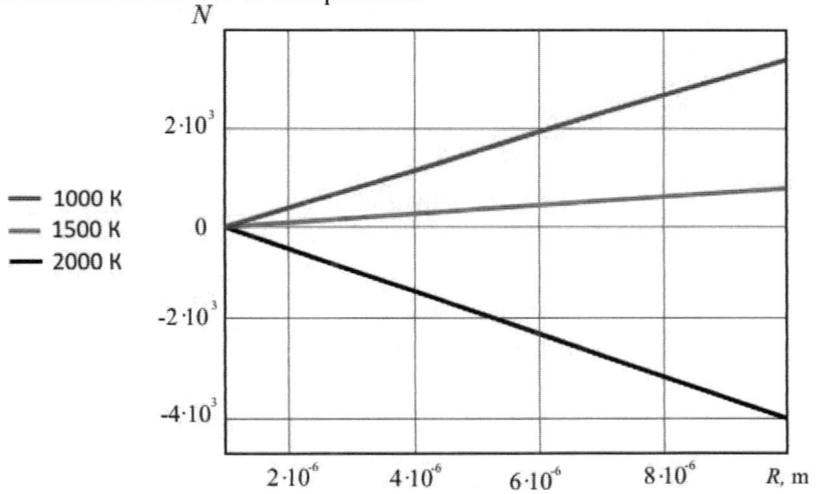

Fig. 7. The dependence of the particle charge on their radius in units of electron charge at different temperatures of the plasma, calculated using the formula (7).

References

[1] Fortov V.E., Khrapak A.G., Yakubov I.T.Physics of strongly coupled plasmas. Textbook. allowance. M. FIZMATLIT , 2004 . 528 .
[2] Dautov G.Yu., Sabitov Sh.R., Fayrushin I.I. 2007 Vestnik of Tupolev KSTU **1** 29-32
[3] Dautov G.Yu., Dautov I.G., Fayrushin I.I. 2009 Vestnik of Tupolev KSTU **1** 57-59.
[4] Dautov I.G., Mardanshin R.M., Fayrushin I.I., Ashrapov T.F. 2010 Vestnik of Tupolev KSTU **3** 143-148.
[5] Dautov I.G., Kashapov N.F., Mardanshin RM, Fayrushin I.I. 2010 Vestnik of Tupolev KSTU **4** 134-136.
[6] V.I. Vishnyakov 2012 Phys. Rev. E **85** 026402.
[7] L.D. Landau , E.M. Lifshitz. Statistical Physics, Moscow, 1976 , 584 pp.

Characteristics of potential well for nondegenerate electronic gas in a equilibrium dusty plasma

G Dautov[1], N Kashapov[2], I Fayrushin[2] and I Dautov[1]

1 Department of General Physics. Kazan National Research Technical University after name A.N.Tupolev – KAI, Kazan, Russian Federation

2 Department of Technical Physics and Energetics. Kazan Federal University, Kazan, Russian Federation

E-mail: fairushin_ilnaz@mail.ru

Abstract. Calculated of the potential well for electrons of the dust-electron plasma depending on the temperature, size of the dust particles, the distance between them and the kind of material of which they consist. The shape of the potential well is significantly different from the rectangular approximation.

1. Introduction

Dusty plasmas in which the concentration of ionized gas atoms is negligible compared to the concentration of electrons, called dust-electron plasma [1-6]. Such a plasma is formed during combustion of various fuels at atmospheric pressure, plasma chemical reactors, MHD generator channels and several processes for obtaining the functional coating with the gas discharge [7,8]. At the temperature of the working gas from 1000 K to 2000 K, solid or liquid dust particles are positively charged as a result of electron emission from the surface and therefore the gas phase contains free electrons . The presence of charged dust particles leads to an uneven distribution of electric potential and the formation of potential wells for electrons.

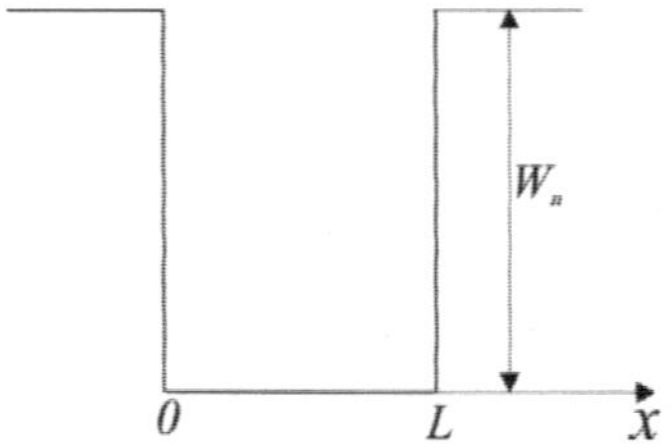

Fig. 1.

Distribution of potential and concentration of electrons inside and around the dust particles depend on the parameters of the potential well. Usually in the approximate calculations believe that the potential well is as shown in Fig. 1, where$L$ – characteristic linear dimension of the solid particle, W_n – the depth of the potential well. It is assumed that in the areas$x<0$ and$x>L$free electron concentration is equal to zero.

We show that this approach has some major drawbacks.Static equilibrium condition of the electron gas can be written as [9]

$$\mu - q\varphi = const, \tag{1}$$

where μ- Fermi energy, q – the absolute value of the electron charge, φ - the electric potential.

We write the Gauss theorem to the electrostatic field

$$\iint_S E_\psi dS = \frac{Q}{\varepsilon\varepsilon_0}, \tag{2}$$

where ε - относительная диэлектрическая проницаемость, ε_0 – relative dielectric permittivity, ε_0 - electric constant, $E_\psi = E\cos\alpha$ - the projection of the electric field on the external normal, Q – charge inside the surface S. If the S take the surface of a dust particle in the form of a sphere of radius R, then equation (2) takes the form

$$4\pi R^2 \varepsilon\varepsilon_0 E(R) = Q, \tag{3}$$

If the electrons are only inside the particles in $r \le R$, space charge Q is zero. Then from (3) that $E(R)=0$.

From (1) we find

$$\frac{d\mu}{dr} - q\frac{d\varphi}{dr} = 0, \tag{4}$$

The Fermi energy [8] of a non-degenerate electron gas is

$$\mu = \theta \cdot \ln\left[\frac{n_e}{2}\left(\frac{h^2}{2\pi m\theta}\right)^{3/2}\right], \tag{5}$$

Substituting this expression in (4) with the communication potential of the field intensity gives

$$\theta\frac{dn_e}{dr} + n_e qE = 0.$$

Taking into account the equation of state for non-degenerate electron gas

$$p = n_e\theta, \tag{6}$$

where $\theta = kT$ - statistical temperature, k – Boltzmann's constant, T – absolute temperature, we find

$$\frac{dp}{dr} + n_e eE = 0. \tag{7}$$

The value $\frac{dp}{dr}$ is the power of pressure, acting on unit volume of the electron gas, and $F = n_e qE$ - the force per unit volume of the electron gas with the electric field. If in the area $r < R$ concentration of electrons is constant and equal n_{e0}, and when $r \ge R$ it is zero, when $r = R$ we have $\frac{dp}{dr} = \infty$ and force $n_e qE(R)$ equal to zero. Thus, equation (7) does not hold. Therefore, the assumption of a constant electron density in the area $r < R$ and the absence of electrons in the area $r > R$ is wrong. To resolve this issue, a closer look at the parameters of the potential well of the potential distribution and the concentration of electrons inside a dust particle in its vicinity.

2. Theoretical background

Consider this issue in the case of spherical particles. Let the inner radius of the solid particle is equal to R_1, outer radius R, in the region of $R_1 \le r \le R$ is equal to the hole concentration ni and the electron gas in the conduction band is not degenerate. We assume that the distance between dust particles is $2l$. We approximately assume that the distribution of the parameters of a spherically symmetric.

We find the distribution of the potential φ (r) and the electron density n_e (r) in the region $0 \le r \le l$ under the following conditions:

$$n_i(r) = \begin{cases} 0, \text{if } r < R_1 \text{ and } r > R \\ const, \text{if } R_1 \le r \le R \end{cases}. \tag{8}$$

$$\varphi'(0)=0,\ \varphi'(l)=0. \tag{9}$$

Gauss's theorem (2) for this case, we write as

$$\oiint_S \varepsilon\varepsilon_0 E_\psi dS = \iiint_V (n_i - n_e)qdV, \tag{10}$$

where V - the volume inside the surface S.

In view of the Boltzmann distribution for the electron gas

$$n_e = n_{e0} e^{\frac{q\varphi}{kT}}, \tag{11}$$

from (10) is obtained by Poisson-Boltzmann equation

$$\frac{1}{r^2}\frac{d}{dr}\left(r^2 \frac{d\varphi}{dr}\right) = \frac{q}{\varepsilon\varepsilon_0}\left(n_{e0} e^{\frac{q\varphi}{kT}} - n_i\right). \tag{12}$$

By introducing the dimensionless variables

$$x = \frac{r}{R},\ \ \phi = \frac{q\varphi}{kT},\ \ \bar{n}_i = \frac{n_i}{n_{e0}},\ \ a^2 = \frac{q^2 R^2 n_{e0}}{kT\varepsilon\varepsilon_0}. \tag{13}$$

from (12) we obtain

$$\frac{1}{x^2}\frac{d}{dx}\left(x^2 \frac{d\phi}{dr}\right) - a^2\left(e^{\phi} - \bar{n}_i\right) = 0.$$

Consider the case where $|\phi| << 1$. We expand the amount of e^{ϕ} Taylor series and only the first two terms, i.e. $e^{\phi} = 1 + \phi$. With this expression we obtain

$$\frac{1}{x^2}\frac{d}{dx}\left(x^2 \frac{d\phi}{dr}\right) = a^2\left(1 + \phi - \bar{n}_i\right). \tag{14}$$

The general solution of (14) is a function

$$\phi(x) = \bar{n}_i - 1 + \frac{Ae^{ax} + Be^{-ax}}{x}. \tag{15}$$

In the area $0 \le x \le \frac{R_1}{R}$ potential distribution under condition $\phi_1(l)=0$ and $\bar{n}_i = 0$ has the form

$$\phi_1(x) = -1 + \frac{e^{ax} - e^{-ax}}{2ax}. \tag{16}$$

In the area $\frac{R_1}{R} \le x \le 1$ potential distribution can be written as

$$\phi_2(x) = \bar{n}_i - 1 + \frac{C_3 e^{ax} + C_4 e^{-ax}}{x}, \tag{17}$$

where $C_3 = \frac{1}{2a} - \frac{R_1}{2aR}\bar{n}_i e^{-\frac{aR_1}{R}}\left(a + \frac{R}{R_1}\right)$, $C_4 = -\frac{1}{2a} - \frac{R_1}{2aR}\bar{n}_i e^{\frac{aR_1}{R}}\left(a - \frac{R}{R_1}\right)$.

The potential distribution $\phi_3(x)$ in the area $1 \le x \le \frac{l}{R}$:

$$\phi_3(x) = -1 + \frac{C_5 e^{ax} + C_6 e^{-ax}}{x}, \tag{18}$$

Constant C_5 and C_6 are determined from the conditions $\phi_2(1) = \phi_3(1), \phi_2'(1) = \phi_3'(1)$. The condition $\varphi'(l)=0$ is used to determine the electron density n_{e0} at $x = 0$, which is included in all previous formulas.

To establish the shape of the potential well we use the following expression that determines the distribution of the potential energy of the electrons from the W coordinate r:

$$W(r) = \varphi(l) - \varphi(r), \tag{19}$$

3. Results

The results of calculation using formula (19) in the case of spherical particles at T=1000 K, $R_1 = 2\cdot 10^{-7}$ m, $R = 10^{-6}$ m, $n_i = 10^{18}$ m^{-3}, $l = 10^{-5}$ m shown in Fig. 2. As can be seen, in $r>R$, W value increases due to escape of electrons from dust particles in the surrounding area.

In the particular case of R_1=0, the resulting formulas are simplified. In particular, for potential distribution, formulas

$$\phi(x)=\frac{e^a(a-1)-\left(\frac{a\lambda-1}{a\lambda+1}\right)e^{a(2\lambda-1)}(a+1)}{2a\left(e^a(a-1)-(a+1)\left(\frac{a\lambda-1}{a\lambda+1}\right)e^{a(2\lambda-1)}+1+\left(\frac{a\lambda-1}{a\lambda+1}\right)e^{2a\lambda}\right)}\left(\frac{e^{ax}-e^{-ax}}{x}-2a\right),\quad \text{at } 0\le x\le 1,$$

$$\phi(x)=-1+\frac{e^a(a-1)+e^{-a}(a+1)}{2a\left(e^a(a-1)-(a+1)\left(\frac{a\lambda-1}{a\lambda+1}\right)e^{a(2\lambda-1)}+1+\left(\frac{a\lambda-1}{a\lambda+1}\right)e^{2a\lambda}\right)}\frac{e^{ax}+\left(\frac{a\lambda-1}{a\lambda+1}\right)e^{a(2\lambda-x)}}{x},\quad \text{at } 1\le x\le\lambda,$$

where $\lambda = l/R$.

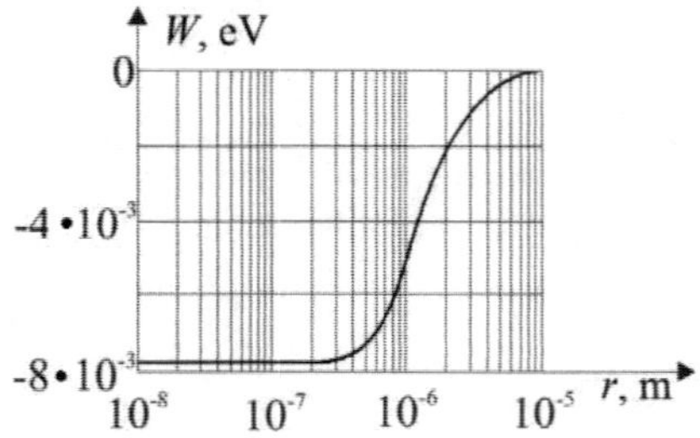

Fig. 2.

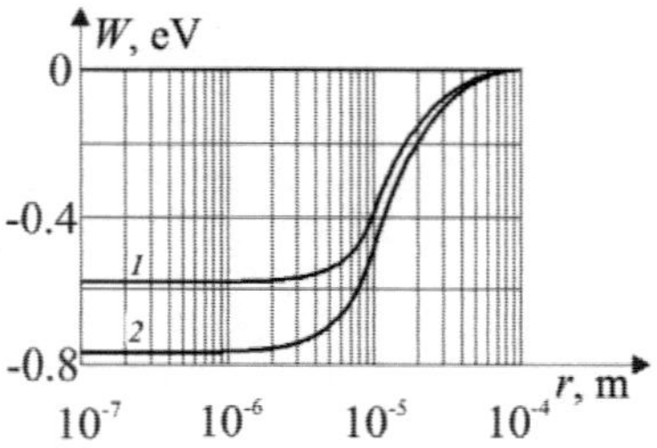

Fig. 3.

The calculation results for the case when the ball at T=1000 K (curve 1) and at T=2000 K (curve 2), $R = 10^{-5}$ m, $n_i = 10^{18}$ m^{-3}, $l = 10^{-4}$ m presented in Fig. 3. As shown, the parameters of the potential wells are temperature dependent. With increasing temperature, the well depth increases. This effect is associated with an increase in electron gas pressure and as a result, high yield of electrons beyond the dust particles with increasing temperature. Thus, it was found that in general, the potential well is not rectangular but has a more complex geometry.

References

[1] Fortov V.E., Khrapak A.G., Yakubov I.T.Physics of strongly coupled plasmas. Textbook. allowance. M. FIZMATLIT , 2004 . 528 .
[2] Dautov G.Yu., Sabitov Sh.R., Fayrushin I.I. 2007 Vestnik of Tupolev KSTU **1** 29-32
[3] Dautov G.Yu., Dautov I.G., Fayrushin I.I. 2009 Vestnik of Tupolev KSTU **1** 57-59.
[4] Dautov I.G., Mardanshin R.M., Fayrushin I.I., Ashrapov T.F. 2010 Vestnik of Tupolev KSTU **3** 143-148.
[5] Dautov I.G., Kashapov N.F., Mardanshin RM, Fayrushin I.I. 2010 Vestnik of Tupolev KSTU **4** 134-136.
[6] V.I. Vishnyakov 2012 Phys. Rev. E **85** 026402.
[7] Solonenko O.P., Gulyaev I.P., Smirnov A. 2008 Technical Physics Letters **24** 22-27
[8] L.D. Landau , E.M. Lifshitz. Statistical Physics, Moscow, 1976 , 584 pp.

Features of radiation gas discharge with liquid electrolyte cathode by using aqueous solutions of sodium chloride

Kh K Tazmeev, I M Arslanov and G Kh Tazmeev

Branch of Kazan Federal University in Naberezhnye Chelny, 68/19 Mira Street, Naberezhnye Chelny, 423810, Russia

E-mail: tazmeevh@mail.ru

Abstract. Experimentally investigated the spectral characteristics of the gas discharge between the liquid electrolyte cathode (aqueous solution of sodium chloride) and a metal anode at high currents (up to 20 A). Found splitting yellow doublet of sodium and the factors affecting the splitting.

1. Introduction

Practical interest in the gas discharge with liquid electrolyte cathode conditioned by the fact that it allows to create a plasma with a significant geometric volume at atmospheric pressure using a comparatively simple technical means. In this regard, it may find application in plasma-technology, in particular in the plasma gasification of waste polymeric materials [1]. For the rational use of plasma gas discharge are required its detailed characteristics in a wide range changes of energy parameters. The objective of this research was to study the spectral characteristics of the gas discharge at high currents.

2. Experiment

The experimental setup is shown in figure 1.

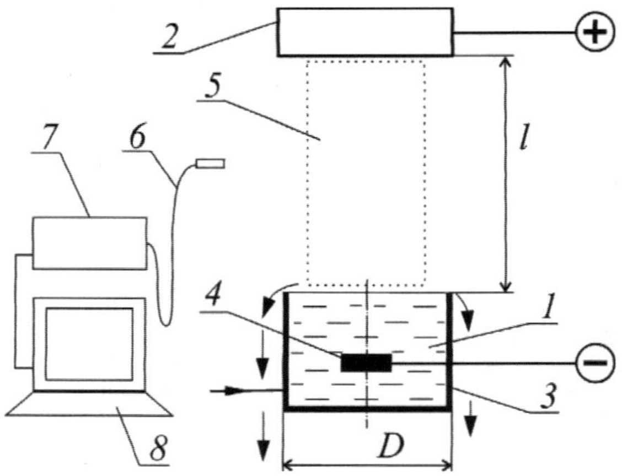

Figure 1. Gas discharge unit. The arrows indicate the direction of movement of the electrolyte.

Journal of Physics: Conference Series **479** (2013) 012015 doi:10.1088/1742-6596/479/1/012015

The discharge was initiated between the liquid electrolyte cathode 1 and the metal anode 2. Electrolyte flowed out of cylindrical passage 3, which has been mounted inside the graphite plate 4 for supplying current. As the electrolyte was used salt solution in distilled water.

The power supply has served three-phase full wave rectifier was connected to the secondary windings of the boosting transformer. Voltage ripple smoothed out by C-L-C-filter. Current is changed by a step variation of ballast resistance.

The radiation spectra were recorded on a high-speed fiber-optic spectrometer AvaSpec-3648 in the wavelength range 484-708 nm with a resolution of 0,15 nm (diffraction grating 1200 lines / mm, the input optical gap is 10 microns). The radiation from the discharge area 5 projected onto to the input optical cable 6 by the collimating lens installed on the receiving end of it. Spectrometer 7 was functionally connected with a computer 8.

3. Results and their discussion

On figure 2 shows the spectrum of the discharge and its fragments amplified in the visible radiation region. On the spectrum shows the leading line of the principal series of sodium (yellow *D*-line). The spectrum can be considered as almost a single line. Compared with the *D*-line intensity of the other components are negligibly small. Against the background of *D*-line, they are indistinguishable. In the mode with the increased integration time (with a CCD detector saturation in the central region of the spectrum) obtained the spectra in which the are observed doublet lines of the sodium atom. Among them, the most intense diffuse series doublet formed by transitions $4^2D_{3/2} \rightarrow 3^2P_{1/2}$ (568,3 nm), $4^2D_{3/2} \rightarrow 3^2P_{3/2}$ (568,8 nm) and $4^2D_{5/2} \rightarrow 3^2P_{3/2}$ (568,8 nm). Doublet relating to the sharp series has a lesser intensity. It consists of lines $5^2S_{1/2} \rightarrow 3^2P_{1/2}$ (615,4 nm) and $5^2S_{1/2} \rightarrow 3^2P_{3/2}$ (616,1 nm). Still less intensity has doublet sodium, consisting of lines 497,9 and 498,3 nm.

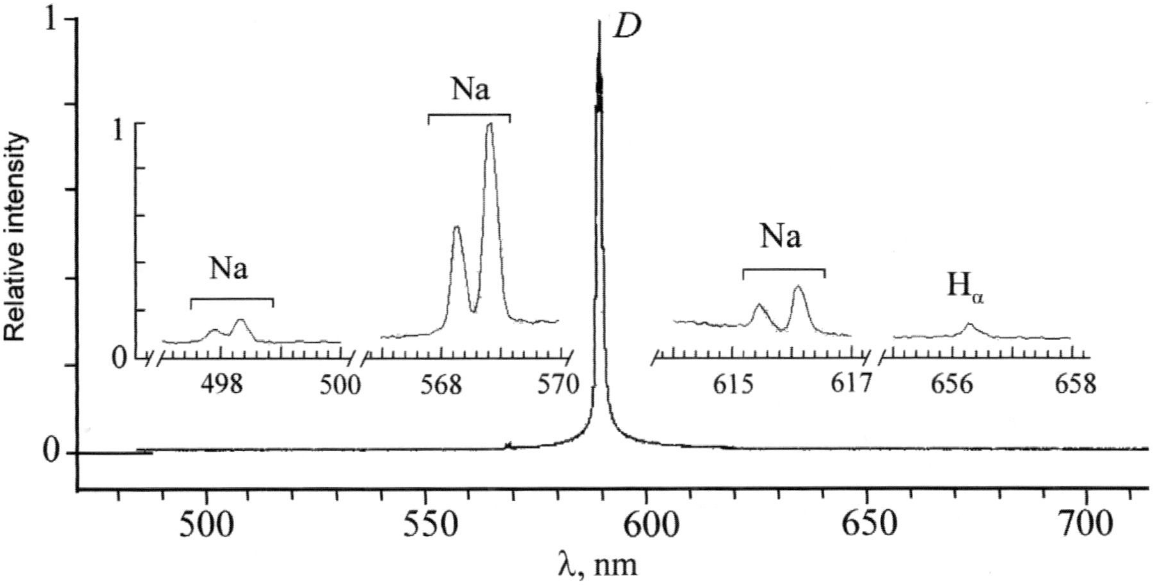

Figure 2. The radiation spectrum of the discharge.

In the strong chunk of spectrum (on the right edge of the bottom of figure 2) clearly stands hydrogen Balmer line H_α (656,3 nm). This line corresponds to the radiative transition from the level with excitation potential 12,088 eV [2]. Therefore, the presence of lines in the spectrum of H_α suggests a presence in the plasma of high-energy electrons that can convert hydrogen atoms from the ground state to the excited state with electron configuration 3*d*.

Exploration of the spectrum in a narrow range of wavelengths indicated that the yellow *D*-line of sodium has a fine structure. According to the Grotrian diagram [2] it should be a doublet (called "yellow sodium doublet"), consisting of lines $3^2P_{1/2} \rightarrow 3^2S_{1/2}$ (λ = 589,6 nm) and $3^2P_{3/2} \rightarrow 3^2S_{1/2}$ (λ = 590,0 nm). In itself observation multiplet splitting yellow sodium *D*-line to form a doublet is quite natural and expected, since the resolution of the spectrometer AvaSpec-3648 used in experimental studies, allows us to see such a result.

Features of radiation, which appeared in the experiments lies in the fact that in a gas discharge except the multiplet splitting of the yellow sodium *D*-line subjected to a supplementary splitting. While the number of components is not permanent and depends on the discharge mode. What is remarkable is that the splitting of spectral lines occurs in the absence of external electric and magnetic fields.

On figure 3 shows the fragments of the spectra obtained at different values of the discharge current. As can be seen, at low currents there is no splitting (figure 3a). It occurs at higher currents (figure 3b) and increases with increasing current (figure 3c). Since additional splitting of spectral lines caused by the nature of the discharge, it can be assumed that it is present and at low currents. Its detection is possible with using a spectral instrument with larger resolving power.

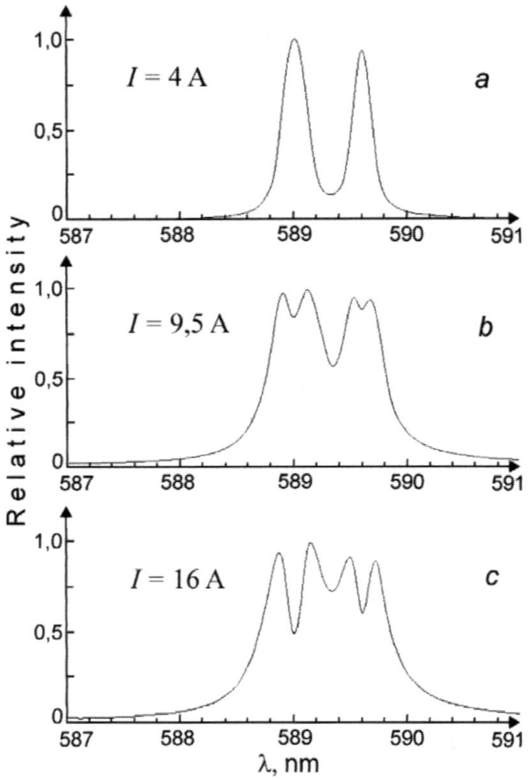

Figure 3. Splitting of the yellow doublet of sodium depending on the discharge current. σ = 10 mSm/sm. l = 7 sm. D = 75 mm. Anode – copper.

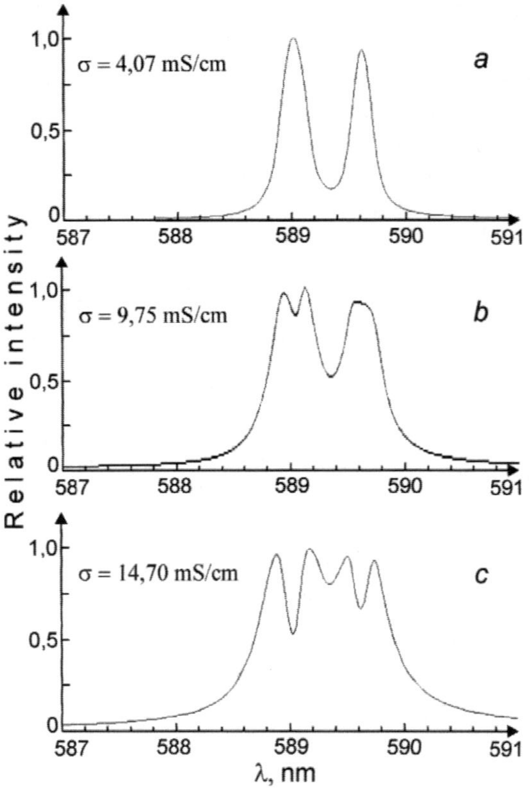

Figure 4. Splitting of the yellow doublet of sodium depending on the electrical conductivity of the electrolyte σ. I = 9,5 A. l = 7 sm. D = 75 mm. Anode – copper.

A similar pattern is observed with an increase in the electrical conductivity of the electrolyte (figure 4). Under the influence of these two factors, improving I and increasing σ, he splitting of the line $3^2P_{1/2} \rightarrow 3^2S_{1/2}$ (λ = 589,6 nm) is greater than the line $3^2P_{3/2} \rightarrow 3^2S_{1/2}$ (λ = 590,0 nm). This is clearly seen in Figures 3b and 4b. It should also be noted that the current increase, and increase the electrical conductivity of the electrolyte lead to increased intensity of the gas discharge. Quantitatively, this effect is confirmed by the expansion of the spectral band in figure 3 and 4.

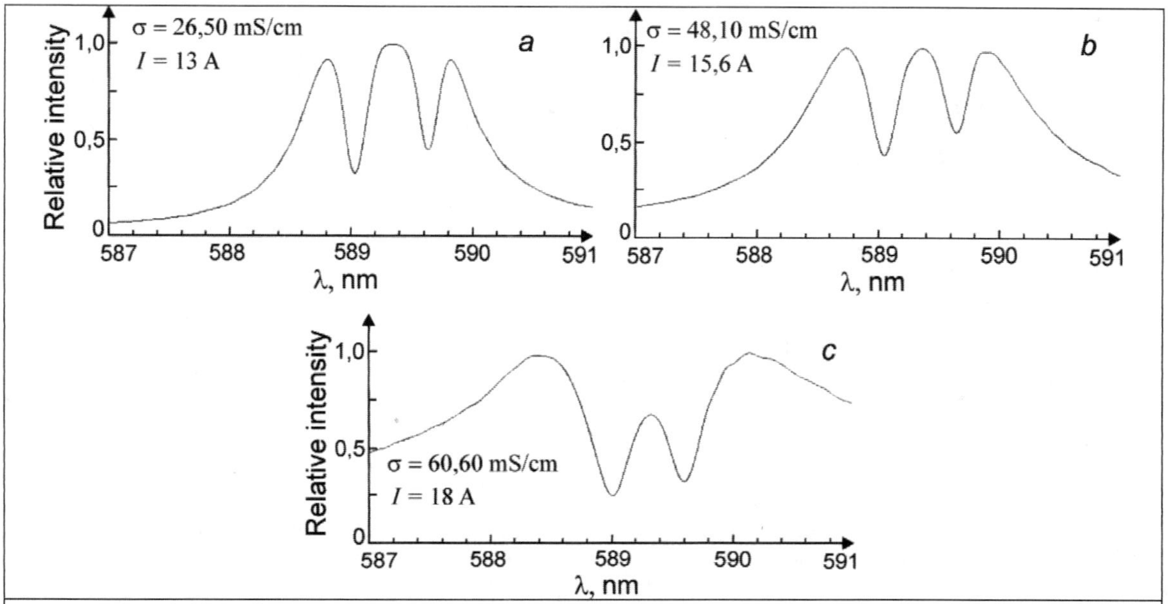

Figure 5. Amplified splitting of yellow sodium doublet *a* and *b* – $l = 5$ sm; *c* – 3 sm. $D = 75$ mm. Anode – grey iron

While simultaneously increasing I and σ an additional splitting of spectral lines of yellow sodium doublet increases (figure 5). It is comparable to the multiplet splitting. Two lines of the spectrum, located in the middle, superimposed on each other. The result is a triplet. The intensity of the triplet midline decreases with the increasing of splitting. Its relative intensity is considerably smaller than the outer lines (figure 5c).

4. Conclusions

In the gas discharge with the liquid cathode electrolyte, when used as electrolyte aqueous solutions of sodium chloride there is a splitting doublet yellow sodium. Increasing the discharge current and the increase of electrical conductivity of the electrolyte lead to increased splitting the spectral lines.

References

[1] Friedland S V, Tazmeev A Kh, Miftakhov M N and Tazmeev Kh K 2006 *Vestnik mashinostroeniya* **7** 72

[2] Babichev A P, Babushkina N A, Bratkovskij A M et al 1991 *Fizicheskie velichiny: Spravochnik* (Moscow: Jenergo atom izdat)

An analytical derivation of non-local dependece of ionazation rate in the glow discharge

A I Sayfutdinov[1]

[1]Department of General Physics. Kazan National Research Technical University after name A.N.Tupolev – KAI, Kazan, Russian Federation

E-mail: s_almaz@ro.ru

Abstract. Based on the analysis of the kinetic equation for the electron distribution function, variables recorded in full energy - coordinate, showed that the rate of ionization is a function of the mean value of the electric field on the electron mean free path in the case of non-uniform electric field in a glow discharge, that is, there is a non-local of the ionization processes. Results are obtained from the numerical solution of non-local dependence of the ionization rate.

1. Introduction

At present, glow discharges is actively used in practical applications, specifically in treating and modifying the surface of materials, in gas lasers, in plasma chemistry, and other purposes.

In the literature, there are a huge number of publications devoted to the glow discharge, and the modeling of its structure. And in most cases, when modeling the structure of the glow discharge in fluid approximation, using the expression for the ionization rate recorded by the Townsend coefficient [1 and other]:

$$Q = \alpha(E) n_e \mu_e E\,, \quad \alpha(E) = Ap\exp\left(-Bp/|E|\right), \tag{1}$$

where n_e, μ_e – the concentration and mobility of electrons, respectively; E – the electric field; A, B – empirical constants depending on the type of gas; p – gas pressure. In this case ionization rate is a function of the local values of the electric field. Such dependence follows from the solution of the kinetic equation for the distribution function of electrons in uniform electric field [1, 2]. However, in the cathode region, the field is highly non-uniform, and degree of non-uniformity can be such that the field change on the free path of a charged particle is comparable to the value of the field. In this case there is no steady drift motion of charges. Moving in a non-uniform electric field, an electron gains energy, that is different from the energy corresponding to the local value of the electric field $E(x)$ and should be at least traversed to them of the potential difference, that is, the ionization processes, should be superior to the equilibrium values corresponding to the local field $E(x)$.

Non-local dependence of ionization processes are often obtained by using different hybrid schemes [2-5]. Non-local model of hollow cathode and glow discharge was obtained in [6].

2. Introduction

To account for the influence of non-uniform electric field on the rate of ionization write a one-dimensional steady-state kinetic equation for the electron distribution function in the two-term and the single-level approximation in variables - full energy and the coordinate [1,7]:

$$\frac{1}{\upsilon}\frac{\partial}{\partial x}\left(\upsilon D(w)\frac{\partial f_0}{\partial x}\right)+\frac{1}{\upsilon}\frac{\partial}{\partial \varepsilon}\left[\upsilon\left(D_\varepsilon(x,w)\frac{\partial f_0}{\partial \varepsilon}+V_\varepsilon(w)f_0\right)\right]- \\ -\nu_{in}(w)f_0(x,\varepsilon)+\sqrt{1+\frac{\Delta w}{w}}\nu_{in}(w+\Delta w)f_0(x,\varepsilon+\Delta w)=0, \tag{2}$$

here f_0 – the isotropic part of the distribution; $\upsilon(w)$ – velocity of the electron; $D_\varepsilon(x,w)$ – the energy diffusion coefficient in the external electric field; V_ε – the energy loss in elastic collisions; ν_{in} – the frequency of inelastic collisions; $D(w)=\upsilon(w)\lambda(w)/3$ – spatial diffusion coefficient for electrons with kinetic energy w; $\lambda(w)$ – the mean free path of electrons. All other functions of w are considered depended on ε, x according to $\varepsilon = w - q_e\varphi(x)$. The first term in (2) describes the spatial diffusion, the second term describes the energy losses associated with elastic collisions, and the third - the energy loss in inelastic collisions, the fourth - the appearance of an electron with kinetic energy due to the loss of their energy in the inelastic process. Analysis of the distribution functions of electrons in inhomogeneous fields done in [8, 9, and others].

We assume that:

1) the cross section of the elementary processes are independent on the energy w, that is, the frequency of elastic and inelastic collisions are defined, respectively $\nu_{el}(w)=n_0\upsilon\sigma_{el}=\sqrt{2w/m}\,n_0\sigma_{el}$, $\nu_{in}(w)=n_0\upsilon\sigma_{in}=\sqrt{2w/m}\,n_0\sigma_{in}$, where n_0 – the concentration of neutral particles, σ_{el} – cross section of elastic collisions, σ_{in} – the cross section of inelastic collisions;

2) the electric field is a linear function in the cathode region of the glow discharge;

3) scale spatial heterogeneity is more than the distance at which the electron acquires the mean energy;

4) the loss of energy by electrons in elastic collisions with neutral gas particles are small compared to the energy losses in inelastic collisions. In this case, we assume that the value of the transmitted energy Δw is comparable to the finite energy of the electron w, i.e., $\Delta w \approx w$.

Based on the above equation (2) can be rewritten as:

$$-\frac{d}{dx}\left(\frac{2w^{3/2}}{3m\nu_{el}}\frac{df_0(x,\varepsilon)}{dx}\right)+\nu_{in}(w)\sqrt{w}f_0(x,\varepsilon)=\sqrt{2w}\,\nu_{in}(w+\Delta w)f_0(x,\varepsilon+\Delta w). \tag{3}$$

The boundary conditions for equation (3) are in the requirement for the limited solutions at $x\to\infty$ and $x\to x_\varepsilon$ where x_ε is determined from the condition $w=\varepsilon+q_e\varphi(x)=0$. Using the Green function $G(x,x')$ of the differential operator of the left side of equation (2) represent the solution in the form of:

$$f_0(x)=\int_{x_\varepsilon}^{\infty}G(x,x')\sqrt{2w}\nu_{in}(w+\Delta w)f_0(x,\varepsilon+\Delta w)dx'. \tag{4}$$

The Green's function is given by $G(x,x')=f_1(x')f_2(x)$, where f_1 and f_2 – solutions of the homogeneous equation

$$-\frac{d}{dx}\left(a(w)\frac{df_0}{dx}\right)+b(w)f_0(x,\varepsilon)=0,$$

with the boundary conditions $x \to x_\varepsilon$ and $x \to \infty$, respectively. Here $a(w) = 2w^{3/2} / 3m\nu_{el}$, $b(w) = \nu_{in}(w)\sqrt{w}$. By demanding the continuity solutions to the variable at the point $x = s$, and the conditions on leap of the derivative, we obtain the following conditions on f_1 and f_2:

$$c_1 f_1(s) = f_2(s) \quad \text{и} \quad \left.\frac{df_2}{dx}\right|_{x=s} - c_1 \left.\frac{df_1}{dx}\right|_{x=s} = \frac{1}{a(w)}. \tag{5}$$

Determine the form of the function. Suppose $\langle w \rangle$ – mean energy of the electron on the mean free path λ_{free}, E^*_{free} – the mean value of the electric field on λ_{free}. Considering that $dw = -q_e E dx$, the expression for the mean energy of an electron can be written as follows $w = \langle w \rangle = q_e E^*_{free}(x - x_\varepsilon)$. Then

$$a(w) = \frac{2w\sqrt{w}}{3m\nu_{el}} = \frac{2w}{3mn_0\sigma_{el}}\sqrt{\frac{m}{2}} = \frac{2q_e E^*_{free}(x - x_\varepsilon)}{3mn_0\sigma_{el}}\sqrt{\frac{m}{2}},$$

$$b(w) = \nu_{in}\sqrt{w} = \sqrt{\frac{2}{m}} n_0 \sigma_{in} w = \sqrt{\frac{2}{m}} n_0 \sigma_{in} q_e E^*_{free}(x - x_\varepsilon).$$

and equation (4) becomes:

$$-\frac{d}{dx}\left((x - x_\varepsilon)\frac{df_0}{dx}\right) + 3n_0^2 \sigma_{in}\sigma_{el}(x - x_\varepsilon) f_0 = 0. \tag{6}$$

It's solution under the condition $x \to x_\varepsilon$ and satisfies the condition (5) is

$$f_1(x) = \sqrt{\frac{m}{2}} \frac{3n_0\sigma_{el}}{q_e E^*_{free}} I(\kappa(x)), \tag{7}$$

where $I(\kappa(x))$ – the Bessel function of imaginary argument, $\kappa(x) = n_0\sqrt{3\sigma_{in}\sigma_{el}E^*_{free}}(x - x_\varepsilon)$.

From the condition that the amount of transmitted energy is comparable in magnitude to the final electron energy at which there is ionization of neutral particles, it follows that the right-hand side of (3) is a function that is localized near the point x_ε. Outside this interval is negligibly small compared to the other terms in (3), i.e.

$$\frac{\sqrt{2w}\,\nu_{in}(w + \Delta w) f_0(x, \varepsilon + \Delta w)}{\sqrt{w}\,\nu_{in}(w) f_0(x, \varepsilon)} << 1. \tag{8}$$

From the last relation follows that the integral in (4) is dialed from the neighborhood, then the asymptotic behavior of the function $f_0(x)$ under the condition $x \to \infty$ can be defined by the expression

$$f_0(x) = f_2(x) \int_{x_\varepsilon}^{\infty} f_1(x')\sqrt{2w}\,\nu_{in}(w + \Delta w) f_0(x, \varepsilon + \Delta w) dx'. \tag{9}$$

In the region where the valid such representation $f_2(x)$ has of a quasi-classical form (Wentzel-Kramers-Brillouin approximation) [7]:

$$f_2(x) \sim \frac{1}{\sqrt{w}} \exp\left(-\int_{x_\varepsilon}^{x} \chi(w) dx''\right), \tag{10}$$

where $\chi(w) = 3m\nu_{el}\nu_{in} / 2w$. Thus, in account of (6) and (9) and introducing the notation

$$C(E_{sv}^{*}) = \int_{x_\varepsilon}^{\infty} \sqrt{\frac{m}{2}} \frac{3n_0\sigma_{el}}{q_e E_{free}^{*}} I(\kappa(x))\sqrt{2w}\, v_{in}(w+\Delta w) f_0(x,\varepsilon+\Delta w) dx',$$

expression (9) takes the form

$$f_0(x) = C(E_{free}^{*}) f_2(x). \tag{11}$$

Taking into account condition 2 the mean value of electric field $E_i^{*}(x,w)$ at the interval $(x_\varepsilon, x) = (x-\Delta x, x)$ corresponding to the length of the path of an electron λ_i (index $i = ion$ – ionization mileage $i = free$ – free path) as follows [8, 11]:

$$E_i^{*}(x,w) = \left(E - \frac{dE}{dx}\frac{\lambda_i}{2}\right). \tag{12}$$

Taking into account $dw = -q_e E dx$ an expression for the distribution function can be rewritten as:

$$f_0(x) = \frac{C\left(E_{free}^{*}\right)}{\sqrt{w}} \exp\left(-\frac{1}{-q_e E_{ion}^{*}} \int_0^{w} \chi(w'')dw''\right). \tag{13}$$

The ionization rate is known to be defined by the equation $Q(x) = \int_I^{\infty} v(w) f_0(w,x) dw$ or with (13)

$$Q(x) = \frac{C\left(E_{free}^{*}\right)}{\sqrt{w}} \exp\left(-\frac{1}{-q_e E_{ion}^{*}} \int_0^{w} \chi(w'')dw''\right) \int_I^{\infty} v_{ion}(w) \frac{f_0(w,x)}{f_{ion}(w_{ion},x)} dw.$$

Denoting, $B = \int_0^{I} (\chi(\varepsilon)/p) d\varepsilon / q_e$ we obtain

$$Q(x) = C\left(E_{free}^{*}\right) \exp\left(-\frac{Bp}{E_{ion}^{*}}\right) \cdot \frac{1}{\sqrt{w}} \int_{w_{ion}}^{\infty} v_{ion}(w) \frac{f_0(w,x)}{f_{ion}(w_{ion},x)} dw. \tag{14}$$

From (14) it is clear that the structure of the expression for the ionization rate in the case of non-uniform electric field is similar to (1) with the difference that in the exponent, the value of the local field is replaced by its mean value over the length of the ionization path of an electron, and in the pre-exponential factor - mean the value of the mean free path of electrons.

3. Appendices

To use the latter expression for practical calculations in the system of equations describing the structure of glow discharge in the fluid approximation it can be rewritten as follows:

$$Q = Apn_e\mu_e \left(E - \frac{dE}{dx}\frac{\lambda_{free}}{2}\right) \exp\left(-\frac{Bp}{\left|E - \frac{dE}{dx}\frac{\lambda_{ion}}{2}\right|}\right). \tag{15}$$

Expression (15) allows us to formulate the problem as a system of differential equations, in contrast to the system of integral-differential allows you to simplify the numerical algorithm for solving equations, taking into account the most important phenomena (non-locality of the ionization processes) occurring in the glow discharge, and using the known experimental data A and B characterizing the type of gas.

The resulted analysis confirms the validity of the applied in [7,10] the expression for the ionization rate as a function of the mean value of the electric field on the electron path length to account for non-locality of the ionization processes in the glow discharge.

References

[1] Raiser Y P 1997 Gas Discharge Physics 538 (Springer–Verlag, New York)
[2] Surendra M, Graves D B, Jellium G M 1990 *Phys. Rev. A* (Vol 41) P 1112
[3] Goto M, Kondon Y 1998 *Jpn. Appl. Phys* (Vol 37) P 308
[4] Fiala A, Pichford L C, Boeuf J P 1994 *Phys. Rev. E.* (Vol 49) P 5607
[5] Donko Z, Hartmann P, Kutasi K 2006 *Plasma Sources Sci. Technol* (Vol 15) P 178
[6] Gorin V V 2010 Non-local model of hollow cathode and glow discharge – theory calculations and experiment comparison *Eur Phys J D* (Vol 59) pp 241–247 DOI: 10.1140/epjd/e2010-00165-9
[7] Kudryavtsev A A, Smirnov A S Tsendin L D 2010 Glow Discharge Physics: the manual (Lan' St. Petersburg)
[8] Dautov G Y, Timerkaev B A 1996 Generators non-equilibrium gas-discharge plasma 1996 (FEN, Kazan)
[9] Tsendin L D 1982 The distribution function of electrons in a weakly ionized plasma in a non-uniform electric fields: the small field, the energy balance is determined quasielastic collisions *Plasma Physics* (vol 8) pp 169 –176
[10] Tsendin L D 1982 The distribution function of electrons in a weakly ionized plasma in a non-uniform electric fields: the large field, the energy balance is determined by inelastic collisions *Plasma Physics* (vol 8) pp 400–409
[11] Saifutdinov A I, Timerkaev B A 2012 Drift model of a glow discharge with account for the nonlocal value of the electric field in the ionization source *Journal of Engineering Physics and Thermophysics* (vol 85) pp 1202 – 1207

Pulsating combustion of gas fuel in the combustion chamber with closed resonant circuit

E V Yallina[1], V M Larionov[1], O V Iovleva[1]
[1]Kazan (Volga region) federal university

E-mail: evgeniya.yallina@gmail.com

Abstract. In the combustion chambers of the pulsation of gas flow oscillation greatly accelerate heat dissipation to the walls of the combustion chamber and improve combustion efficiency as compared with a uniform combustion mode. This allows you to effectively solve a number of problems of industrial power, including an environmentally friendly combustion products. Significant drawback of such systems - the emitted noise exceeding the permissible requirements. One solution to this problem - the separation of the resonance tube into 2 parts connected at the output to the interference of sound waves. The results of theoretical studies pulsating combustion technical mixture of propane in the system, consisting of a combustion chamber and two resonance tubes forming a closed resonant circuit. Resonators have a variable length. Calculations have shown that under certain oscillation of the resonator length to the first resonant frequency of the system is achieved by reducing SPL more than 15 dB. For oscillations at a second resonant frequency is the complete elimination of noise while maintaining intense oscillations in the combustion chamber.

1.Introduction

In this paper we consider the system with two resonance tubes of different lengths. In this case, the path length of sound waves in the resonator is different, and it may happen that pressure pulsations of the outlet tubes have opposite phase. When the pipe ends have a common outlet for the combustion products, the sound waves emitted will interfere. Condition, when the path difference of the sound waves will be equal to half the wavelength, it is possible to achieve a minimum of intensity of sound holes required for combustion products. Furthermore, the use of two tubes will increase heat-transfer surface, and further intensify the heat exchange process in heating installations pulsating combustion. The aim of this work - the development of a mathematical model of the process of excitation of oscillations in the combustion chamber of gaseous fuel with two resonance pipes and assess the possibility of reducing the level of noise emitted by the system under study .
Fig . 1 is a schematic diagram of study unit.

2. Relations defining the conditions of excitation, frequency and amplitude of the oscillations gas when resonance tube are connected to the outputs

Figure 1 is a schematic diagram of the study unit.

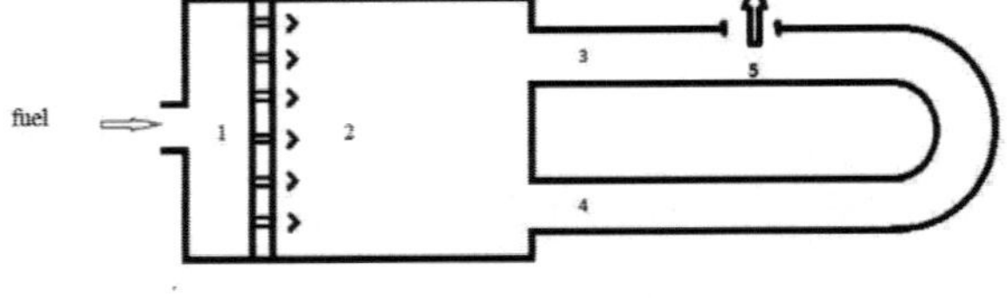

Fig. 1 Schematic diagram of the study unit.
1 - Multi-channel burner
2 - combustion chamber,
3, 4 - the resonance tube
5 - opening for combustion products.

Pulsation rate and pressure in the combustion chamber and the resonance pipes described known [1] expressions.

$$v_k'(x_k,t)=C_k e^{i\omega t}\left(1-\frac{bx_k}{a}\right)^{1/2}\cos(\varphi_k-\psi_k),\ \psi_k=\frac{\omega\beta}{b}\ln\left(1-\frac{bx_k}{a}\right), \quad (1)$$

$$p_k'(x_k,t)=-i\rho_k c_k C_k e^{i\omega t}\left(1-\frac{bx_k}{a}\right)^{1/2}\left[\frac{b}{2\omega}\cos(\varphi_k-\psi_k)+\beta\sin(\varphi_k-\psi_k)\right], \quad (2)$$

where the index k is 2, 3, 4, $c_k=(a-bx_k)$ - speed of sound, φ – angle depending on the boundary conditions at the ends of the channel, a - a linear coefficient of sound velocity distribution, b – the gradient of the speed of sound in the hot gas, ρ – gas density, $\beta=[1-(b/(2\omega))^2]^{1/2}$ parameter in the solution of the wave equation; C – the maximum amplitude of vibration velocity, ω – angle (cyclic) frequency.

At the entrance to the resonance tubes ($x_3=0,\ x_4=0$) boundary conditions after the linearization of the equation of conservation of mass and momentum take the following form:

$$(S_3/S_4)\,v_{3,0}'+v_{4,0}'=(S_c/S_4)\,ip_{3,0}'/Y_{2,l},\ p_{3,0}'=p_{4,0}', \quad (3)$$

where $Y_{2,l}$ - The output impedance of the combustion chamber [2]. At the exit of the conditions of the resonance tubes ^

$$S_3v_3'(l_3,t)+S_4v_4'(l_4,t)=S_lv_l',\ p_{3,l}'=p_{4,l}'=p_l'. \quad (4)$$

Believe $p_l'=iY_l\vartheta_l'$, obtain $(S_3/S_4)\,v_{3,l}'+v_{4,l}'=(S_l/S_4)\,p_{3,l}'/(iY_l)$, $p_{3,l}'=p_{4,l}'$ If the resonance tubes have a common outlet opening, then:

$$Y_l=\rho_l c_l \mathrm{tg}\,(\omega l_l^*/c_l),\ l_l^*=l_l+0{,}61R_l.$$

We denote the $S_3/S_4=\sigma_{3,4}$; $\rho_{2,l}c_{2,l}(S_c/S_4)/Y_{2,l}=\sigma_0(\omega)$, $\rho_{3,l}c_{3,l}(S_l/S_4)/Y_l=\sigma_l$.

From the boundary conditions (3), (4) taking into account the expressions (1), (2) we obtain the following relations:

$$\varepsilon_{3,4}C_3\cos\varphi_3+C_4\cos\varphi_4=\sigma_0C_3\left(\frac{b}{2\omega}\cos\varphi_3+\beta\sin\varphi_3\right),$$

$$C_3\left(\frac{b}{2\omega}\cos\varphi_3+\beta\sin\varphi_3\right)=C_4\left(\frac{b}{2\omega}\cos\varphi_4+\beta\sin\varphi_4\right),$$

$$(\varepsilon_{3,4}\lambda_3C_3\cos(\varphi_3-\psi_{3,l})+\lambda_4C_4\cos(\varphi_4-\psi_{4,l})=-\sigma_l\lambda_3C_3\left[\frac{b}{2\omega}\cos(\varphi_3-\psi_{3,l})+\beta\sin(\varphi_3-\psi_{3,l})\right],$$

$$\rho_{3,l}c_{3,l}\lambda_3C_3\left[\frac{b}{2\omega}\cos(\varphi_3-\psi_{3,l})+\beta\sin(\varphi_3-\psi_{3,l})\right]=\rho_{4,l}c_{4,l}\lambda_4C_4\left[\frac{b}{2\omega}\cos(\varphi_4-\psi_{4,l})+\beta\sin(\varphi_4-\psi_{4,l})\right],\ \text{где}\ \lambda_k=\sqrt{1-(bl_k/a)}$$

The condition of uniqueness of the solution of these systems of equations for the unknown C_3, φ_3, C_4, φ_4 allows you to find the frequency of oscillation of gas. The formula for calculating the amplitude of the pressure oscillations in the combustion chamber was the energy method [1] has the form:

$$p_c=\frac{a_c^{(L)}-a_{W,c}-a_{W,3}-a_{W,4}-a_l^{(L)}}{a_c^{(N)}+a_l^{(N)}},$$

where a_c, a_w, a_l – the dependency ratio A_c, A_w, A_l the amplitude of the pressure fluctuations in the plane of the heat input, the indices L and N represent linear and non-linear approximation, respectively.

3. Calculation of oscillation gas depending on the geometrical parameters of the installation

Were calculated amplitudes of velocity fluctuation of the gas outlet pipe resonance depending on the length of one of the tubes at a constant length of a resonant circuit. These data are given the phase difference of velocity fluctuations in the gas pipe ends are shown in Table. 1. The calculation results show that, depending on the location of the outlet in the resonant circuit of the gas velocity pulsations are the same or the opposite phase when the value $U_{l,4}$becomes negative.Comparison of tabular data with the results shown in Fig. 2 shows that the minimum value of the sound pressure level in the outlet corresponds to the condition where the gas velocity fluctuations at the ends of the resonance pipes have the same amplitude and opposite phase.

Table 2. Dependence frequency oscillations excited gas ultrasound, gas velocities at the ends of the resonance tubes and the level of noise emitted from a single length of pipe resonance circuit at a constant length, L = 0,5. (secondharmonic)

l_3	f, Hz	p_1, dB	U_{l3}, м/с	U_{l4}, м/с	p_L, dB
0,05	862	129,0	0,50	-0,24	75,5
0,06	847	129,5	0,49	-0,29	72,5
0,08	838	129,5	0,40	-0,33	61,5
0,09	840	129,0	0,34	-0,34	39, 5
0,091	841	129,0	0,34	-0,34	3,5
0,092	841	129,5	0,33	-0,33	39,5
0,095	843	128,5	0,31	-0,33	51,0
0,1	847	128,0	0,29	-0,33	57,50
0,12	868	125,0	0,18	-0,27	65,00
0,14	899	120,0	0,09	-0,18	65,0
0,16	937	105,5	0,02	-0,04	53,5

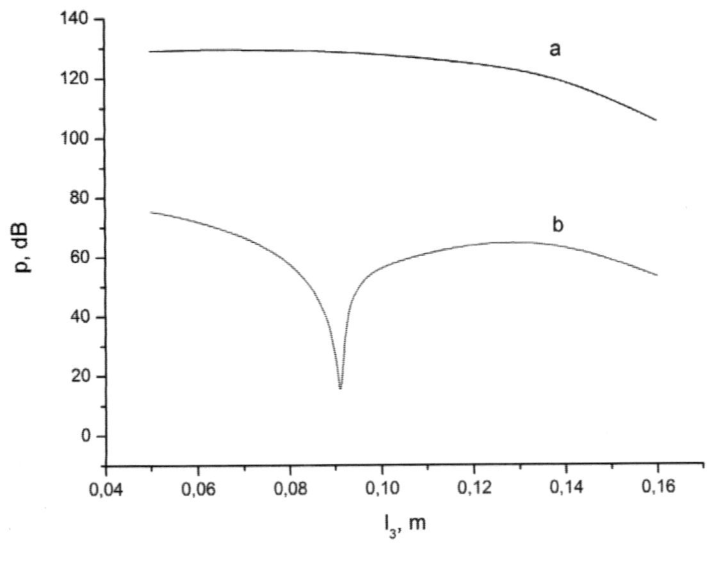

Fig. 1 The dependence of the sound pressure level on the length of one of the resonance tube at a constant contour length L = 0,5 m;

4. Conclusions

1) The mathematical model and method of calculation of the excitation conditions and parameters of gas oscillations in the combustion chamber of the pulsation of fuel with two resonance tubes , which form a closed loop.

2) Found that for a fixed length of one of the resonators , extending the second resonator can be achieved alternating excitation of oscillations of the first gas and the second frequency system.

3) It was found that at an elongation of one of the resonators with respect to another can be achieved by phase transition to the gas velocity pulsations antiphase output of the resonators and reduce the noise by more than 15 dB for a first oscillation frequency of the system .

4) Complete elimination of the noise occurs at the excitation oscillations of the gas with the second frequency system when the pulse rate of the gas from the resonance tubes have the same amplitude but opposite phase.

5) When designing industrial installations denoised recommended geometric parameters of the combustion chamber and resonance tube set such that the excited vibration with the frequency of the second system , and the outlet was located at the resonant circuit at a point to which one longer than the other part of the circuit by an amount equal to half the length sound wave.

References

[1] LarionovVM, ZaripovRG 2003Self-oscillations of gas in installations with combustion 237 (KSTU, Kazan)

[2] LarionovVM, Beloded O V 2006Vibrationcombustion in a tube with jump of cross-section*Izv. VUZ.AviatsionnayaTekhnika (vol 1)*pp 30–33

Journal of Physics: Conference Series **479** (2013) 012018 doi:10.1088/1742-6596/479/1/012018

Technological peculiarities of deposition anti-reflective layers in low-e coatings

Y Yurjev[1] and D Sidelev[2]

1 Head of Laboratory, Department of Hydrogen Energy and Plasma Engineering, National Research Tomsk Polytechnic University, Tomsk, Russian Federation

2 Department of Hydrogen Energy and Plasma Engineering, National Research Tomsk Polytechnic University, Tomsk, Russian Federation

E-mail: Yurjev@tpu.ru

Abstract. This article reports on the investigation of technological features magnetron sputtering for deposition anti-reflection layers in low-emission (low-e) coatings. The three-layer TiO_2-Cu-TiO_2 films were deposited by dual and planar magnetron sputtering systems (MS) on glass substrate. Studies of the current-voltage characteristics (CVC) and the hysteresis effect show that deposition of anti-reflection layers is possible in the transition mode with higher rates. For planar magnetron, the stability of electrical discharge parameters is achieved at 60 % O_2 content in mixture. The calculations optical band gap E_g show that anti-reflective films have a rutile or anatase phases that depending on the content O_2 in gas mixture. The optimum deposition conditions of TiO_2 films were determined for all modifications of magnetrons. Anti-reflective layers, which are deposited by balanced dual MS, improve the transparency of low-e coatings (integral T_{VIS} increase in 15%).

1. Introduction

At present, low-e coatings are widely used to reduce heat loss in residential buildings [1]. The simplest structure of such multilayer film composition is dielectric-metal-dielectric having the following properties: a high transmittance T_{VIS} in the visible region (approximately 0.8), and high reflection R_{IR} in the infrared range (0.85-0.9) [2]. The main functional layer is a semi-transparent metal (Cu, Ag). The translucence of low-e coating is achieved by deposition anti-reflection layers. Predominantly, such films are oxides metals (TiO_2, SnO_2, ZnO).

The main problem of heat-reflective coating is in the deposition of high-quality anti-reflective layers on account of variability of process parameters in the reactive process. Magnetron sputtering, preferably used for deposition optical coatings, is associated with the following disadvantages: no stability of electrical discharge parameters, poisoning target, electric breakdowns on target [3]. Therefore, the main purposes of this study are to determine the technological features of the magnetron in case of reactive process and investigate the effect of plasma source to the properties of the low-e films. In this work, we used Cu as a model metal material, anti-reflective layer - TiO_2.

2. Experimental work

The low-e coatings were deposited on glass substrate (75x25x1 mm^3) by an ion-plasma installation «Yashma», which is based on diffusion pump AVP-250. The vacuum chamber was pumped down to a

Journal of Physics: Conference Series **479** (2013) 012018 doi:10.1088/1742-6596/479/1/012018

base pressure p_0 (10^{-2} Pa). For cleaning surface, substrates were treated by means of an ion source with closed electron drift. TiO_2 films were reactively prepared by dual and planar magnetron sputtering systems in Ar + O_2 mixture at the constant total pressure 0.12 Pa. The deposition of functional layer (Cu film) was occurred by planar magnetron sputtering systems in Ar at the total pressure $8 \cdot 10^{-2}$ Pa. The technical features of magnetrons are shown in table 1.

To provide the uniformity of the film thickness, substrates moved under plasma sources. For this reason, for evaluation a deposition rate we use such technological term as «nm/pass», which means the thickness of the deposited material in one pass under plasma source. Measurements and control of deposition rate ensured by means of a quartz gauge «Micron-5». The investigations of transparency were by a spectrophotometer SF-2000 in the visible region (400-750 nm). The determination of refraction spectra was produced by the ellipsometer Ellipse-1891. The structure of the TiO_2 films was characterized by X-ray diffraction using Shimadzu XRD-7000S in the Bragg-Brentano configuration.

Table 1. The configuration of magnetrons.

magnetron	power supply	impulse form	repetition frequency, kHz	target size, mm^2	magnetic field
planar	unipolar	rectangle	166	100x700	balanced
dual	bipolar	trapezium	66	94x200	balanced, unbalanced

3. Results and discussion

3.1. The deposition Cu layer

It is well known that high quality low-e coatings reflect infrared radiation (R_{IR}) and transmit in the visible light region (T_{VIS}). Therefore, the thickness of functional layer determined based on the comparison of the R_{IR} and T_{VIS}. According to [1], the formation of a continuous Cu film takes place at least a 9 nm thickness. For deposition Cu layer, work mode of planar magnetron (U=360 V, I=3 A, W=1.1 kW, v_{Cu}=9 nm/pass) was used.

3.2. Peculiarities of deposition TiO_2 films

Magnetron sputtering in a reactive atmosphere is difficult due to a problem of «disappearing anode», a hysteresis of discharge parameters, a drop in the deposition rate. It leads to change of structure films and their optical characteristics. Thence, deposition mode of TiO_2 films determined by the stability of discharge parameters and maximum deposition rate (v_{TiO2}) throughout the reactive process.

The current-voltage characteristics (figure 1) and hysteresis of the electrical discharge parameters were studied (figure 2). For reactive deposition of TiO_2 films, S-shaped current-voltage characteristic is a typical. In the experiment, CVC is essentially linear for three modifications of magnetrons. Moreover, a region of negative dynamic resistance of the discharge (negative slope of the curve) is missing. Such peculiarities were associated with the maintenance of a constant total pressure by gradually reducing the Ar flow in proportion to addition the O_2 flow. This leads to what is we forcedly oxidizes target by reducing intensity of Ar^+ bombardment.

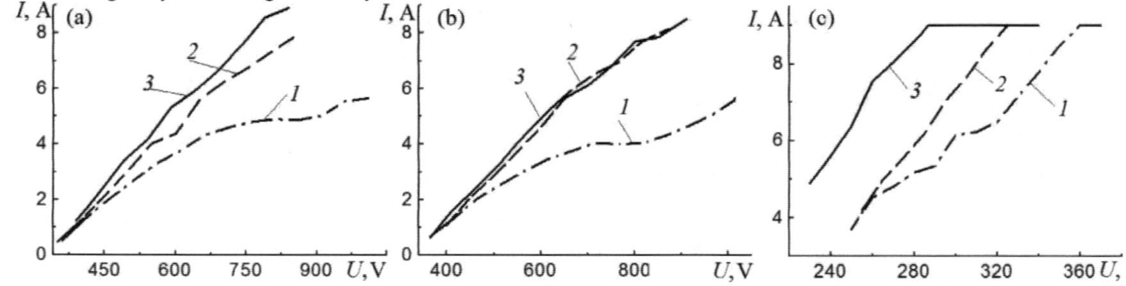

Figure 1. CVC of the discharge: a - balanced dual MS, b - unbalanced dual MS, c - planar MS, 1 - 5% O_2 in a mixture, 2 - 20% O_2 in a mixture, 3 - 60% O_2 in a mixture.

The differences in current and voltage values of CVC for various O_2 content in mixture determined by the degree of target oxidation [4]. For unbalanced dual magnetron, the proximity of the electrical discharge parameters is due to increasing ion current on target. This behavior of the discharge leads to expand of transient region mode. It is presented in more detail in [5]. For planar magnetron, the strong dependence CVC on the O_2 content is specified by a small area of transition mode.

In the experiment, it was found that hysteresis of discharge parameters is completely absent for deposition TiO_2 films by means of the balanced dual magnetron, for other types – hysteresis loop areas are more. It is important to note that the peculiar sharp changes of voltage for planar magnetron aren't observed. This allows for the deposition of stoichiometric TiO_2 films in the transition region of the discharge.

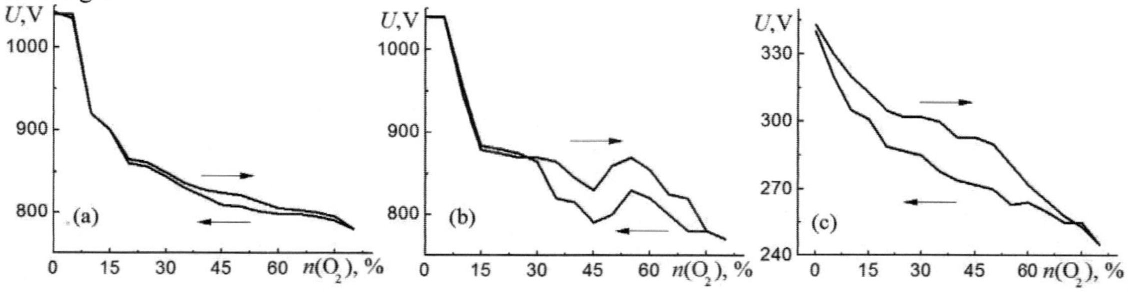

Figure 2. Hysteresis of the operating discharge parameters: a - balanced dual MS, b - unbalanced dual MS, c - planar MS. Arrows indicate the direction of changes in the inlet gas stream O_2.

3.3. Optimal deposition modes TiO_2 layers

For extended periods, attempts to stabilize the operation parameters of planar magnetron in transition mode failed. Magnetron inevitably changed over to sputter fully oxidized target that doesn't occur in the case of using dual magnetrons. As a result, the deposition TiO_2 films by planar magnetron is possible in case of high content O_2 in gas mixture (60%). Therefore, optimal modes were determined by the achievement of maximum deposition rate (table 2). For unbalanced dual magnetron the deposition rate is higher owing to increasing ion current on target.

Table 2. The optimal deposition modes TiO_2 layers.

	Balanced dual MS	Unbalanced dual MS	Planar MS
U, V	830	890	340
I, A	7.5	8	9
W, kW	4.4	5	2.8
v_{TiO2}, nm/pass	3.1	3.8	0,7
$v_{TiO2}/(S_{target} \cdot W)$, nm/(pass·W·cm^2)	$3.75 \cdot 10^{-6}$	$4.04 \cdot 10^{-6}$	$3.57 \cdot 10^{-7}$

Significant numerical differences in the discharge parameters for dual and planar magnetrons are due to their geometric dimensions. Furthermore, the discrepancy of measurements of the electrical characteristics of the dual MS is due to the complex shape of the current impulse. Thence, calculations of deposition rate on power density show that dual magnetrons are more profitable.

The thickness of anti-reflection layers TiO_2 films was determined by means of mathematical modeling of three-layer film in the program IMD [6]. The best translucence is reached at 50 nm for the upper and lower layers. This result correlates with [1].

3.4. Characterization of low-e coating

Transmission spectra of low-e coatings TiO_2 (50 nm)-Cu (9nm)-TiO_2 (50 nm) are shown in figure 3. The lowest transmittance (T_{VIS}) in case using planar MS is associated with lower refractive index n of anti-reflective layers in the visible region. For dual MS, the higher values of T_{VIS} are due to changing film microstructure.

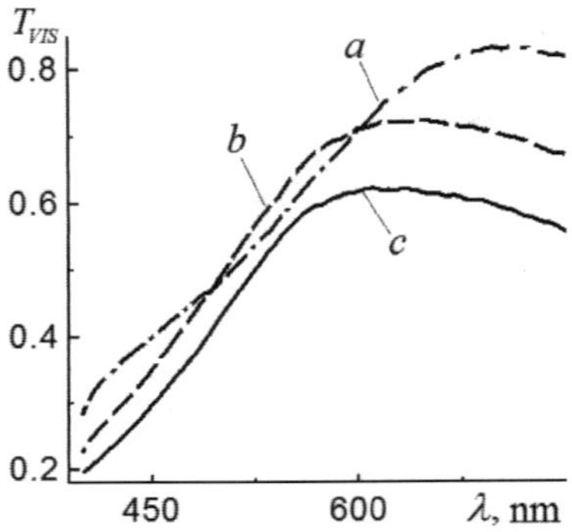

Figure 3. The transmission spectra of low-e coating: a - balanced dual MS, b - unbalanced dual MS, c - planar MS.

According results of XRD, TiO_2 films are X-ray amorphous. For this reason, the investigation microstructure is based on refractive indexes n and optical band gap E_g of TiO_2 films. The E_g calculations were carried out by J. Tauc method, described in [7]. According to table 3, properties of anti-reflective films differ for planar and dual MS due to the selected deposition mode.

It is well known that TiO_2 films, which are reactively prepared at low O_2 content (transition mode), have a rutile phase, for high O_2 content (oxide mode) - an anatase phase. Thus, we assume that anti-reflective films, deposited by dual MS, are more similar to rutile phase, deposited by planar MS - to anatase phase.

Table 3. Characterization TiO_2 films.

	planar MS	dual MS	
		balanced	unbalanced
n at 632.8 nm	1.95	2.22	2.34
E_g, eV	3.52	3.41	3.33

The deposition TiO_2 films with different structure provide a distinction of integral transmittance of low-e coatings. For unbalanced dual MS, the reduction integral T_{VIS} apparently associated with structural changes at the boundary between Cu and upper TiO_2 layers. This aspect requires more detailed study.

4. Conclusions

Main results of our investigation can be summarized as follows:

- the stabilization of electrical discharge parameters of dual MS is possible even in transient mode;
- unbalanced dual MS have a higher width of transient mode;
- dual MS provide higher deposition rates of TiO_2 films concerning planar MS;
- TiO_2 films, which reactively deposited at low O_2 content, have higher n and E_g that associated with content of rutile phase;
- anti-reflective layers, which are deposited by balanced dual MS, improve the transparency of low-e coatings: integral T_{VIS} increase in 15%.

References

[1] Zaharov A, Kovsharov N, Oskomov K, Rabotkin S, Soloviev A and Sochugov N 2012 *Persectivnie Materials* **2** 62-70
[2] Grosse P, Hertling R and Muggenbeurg T 1997 *J. Non-Crystalline Solids* **218** 38-43
[3] Sproul W, Christie D and Garter D 2005 *Thin Solid Films* **495** 1-17
[4] Baroch P, Musil J, Vlcek J, Nam K and Han J 2005 *Surf. Coat. Tech.* **193** 107-111
[5] Sidelev D and Yurjev Y 2013 Titanium Dioxide TiO_2 Thin Films as a Multifunctional Coating *11th Conf. Gas Discharge Plasmas and Their Applications* (Tomsk: Izv. Vuz. Phys.)
[6] Windt D 1998 *Computers in Physics* **12** 360-370
[7] Yurjev Y, Sidelev D and Krivobokov V 2013 *Izv. Vuz. Phys.* **56** 350-354

Abnormal glow discharge in crossed electric and magnetic fields in the presence of reactive gas

A G Luchkin, N F Kashapov and G S Luchkin
Kazan (Volga Region) Federal University, 18 Kremljovskaja str., Kazan 420008, Russian Federation

E-mail: AGLuchkin@kpfu.ru

Abstract. The results of investigations of the low-temperature plasma at process of reactive magnetron sputtering of Ti target in the presence of reactive gas are described. Discharge volt-ampere characteristics for different schemas of reactive gas input are build. Optimal regimes of making strengthening coatings are defined. TiO_x ($0<x<2$) strengthening coatings at plastics are made. Coatings showed high consumer qualities.

1. Introduction

Low-temperature plasma is widely used for making functional coatings. One of the most popular coating methods is magnetron sputtering [1-3]. This method is applying for making different multilayer functional coatings (strengthening, low emission, antireflection, etc.) at wide range of materials: glass, plastic, ceramic, metal [4]. Their structure could consider oxide layers. Such oxide layers could be coated by magnetron sputtering of metallic target in the presence of reactive gas – oxygen. Process of coating at plastic substrate is difficult because of low plastic melding temperature, consequently temperature regime of coating should be chosen correctly [5].

The aim is to make strengthening coatings on plastic by low temperature plasma of magnetron discharge.

2. Experimental plant and methods

Functional schema of experimental plant is showed at pic. 1. It consists of: vacuum chamber 1, door 2, pump system 3, two magnetrons 4, power supply of magnetron and source of ions 5, system of gas input, system of photometric control of film thickness that include spectrophometer S-100, moving system 8 of substrates 9, heating system 10, windows 11, valve for atmosphere air 12, filled-system transducer 13, cooling system 14, ion source 15 [6] .

There is control stand for all systems. Pumping system includes mechanical pump 2NVR90B and diffusion pump ND-200 and provides vacuum at chamber about 10^{-3} Pa. Pressure registration system consists of vacuum-gauges VT1-4 (thermocouple method) and VMB1-2 (ionization method). Gas input in chamber realized by leak valves VTT02.01.31.000 and controller RRG-8. Gas flow counted by constant volume method [7]. Spatial distribution of temperature in region of magnetron measured by probe method at different distances from magnetron.

Volt-ampere characteristics measured within input of work and reactive gases in vacuum chamber at different ratio. Voltage fixed at different values of discharge current. Experimental plant provides gas input in region of magnetron target (schema "A") and in region of substrate (schema "B").

Journal of Physics: Conference Series **479** (2013) 012019 doi:10.1088/1742-6596/479/1/012019

In case of schema "A" reactive gas atoms concentration increases near target surface and probability of their condensation at target is more. In case of schema "B" oxygen atoms concentration near target magnetron is low, consequently, probability of their condensation at target is lower. Oxide appears at target surface because of its high temperature that needs for chemical reaction of oxidation. Target surface oxidation depends on quantity of adsorbed reactive gas atoms, so in case of schema "A" the process is more intensive.

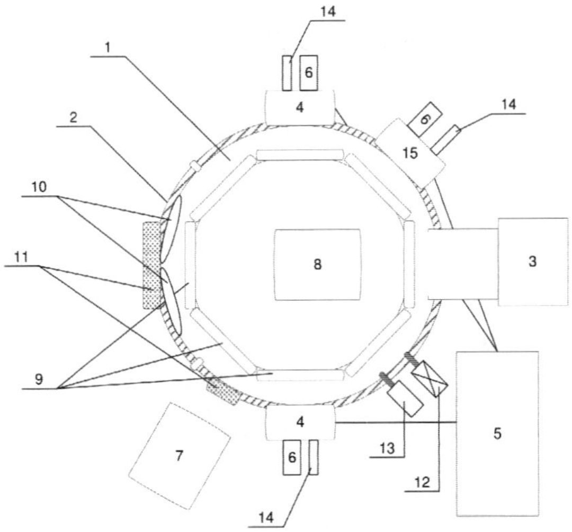

Pic. 1. Functional schema of experimental plant: vacuum chamber 1, door 2, pump system 3, two magnetrons 4, power supply of magnetron and source of ions 5, system of gas input, system of photometric control of film thickness that include spectrophometer S-100, moving system 8 of substrates 9, heating system 10, windows 11, valve for atmosphere air 12, filled-system transducer 13, cooling system 14, ion source 15.

Discharge parameters: cathode voltage 350 V; discharge current 8 A; vacuum in chamber 0,1-0,4 Pa. Target material – Ti. Work gas (Argon) flow was 0,4 – 1,2 mg/sec, reactive gas (Oxygen) flow was 0,2 – 0,7 mg/sec. Sputtering time was 10-30 minutes. Substrates: polycarbonate (PC) and polyethylene-terephthalate (PET). Substrate thickness: 3,0 and 0,45 mm accordingly. As witness was optical glass K-8. There was no additional heating of substrates. Before coating substrates were cleaned by ions from ion source [8]. Film thickness was from 50 to 150 nm.

Titanium oxide films at polycarbonate substrates strength characteristics were measured by micro durometer PMT-3M. Titanium oxide films at polycarbonate substrates strength characteristics were measured by SM-55 according to OST 3-1903-95 [9]. Also two-dimensional method for integral characteristics measuring was used [10]. This method defines film's mechanical characteristics in system "coating-substrate" [11].

3. Results and discussions

Temperature spatial distribution in vacuum chamber at region of magnetron is shown at pic. 2. For making coatings on plastics optimal distance is 10 sm, because of low and equal heating and no film destruction. At the same time maximal coating speed is provided.

At reactive magnetron sputtering of metallic target with direct current reactive gas addition leads to changes of discharge volt-ampere characteristics. There are volt-ampere characteristics at pic. 3 for sputtering only in Argon and in mixture of Argon and Oxygen at different gas input schemas.

Oxide film at target surface appears at work and reactive gas mixture input in chamber. When discharge is burning the transparent coating is making. Increasing of discharge power at defined moment oxide film at target surface sputters totally and metallic coating is making. Discharge power decreasing results to opposite effect but at lower value. The same hysteresis was showed in [12].

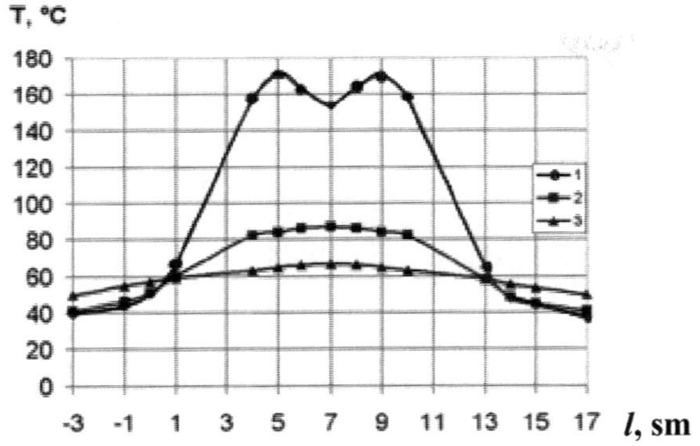

Pic. 2. Temperature spatial distribution in vacuum chamber at region of magnetron. 1 – distance *l* between magnetron target and substrate is 6 sm, 2 – 12 sm, 3 – 18 sm.

At this graphics lower part of hysteresis describes sputtering of oxide film from target surface and making transparent coating on substrate. Higher part of hysteresis describes sputtering of metal target and making metallic coating on substrate.

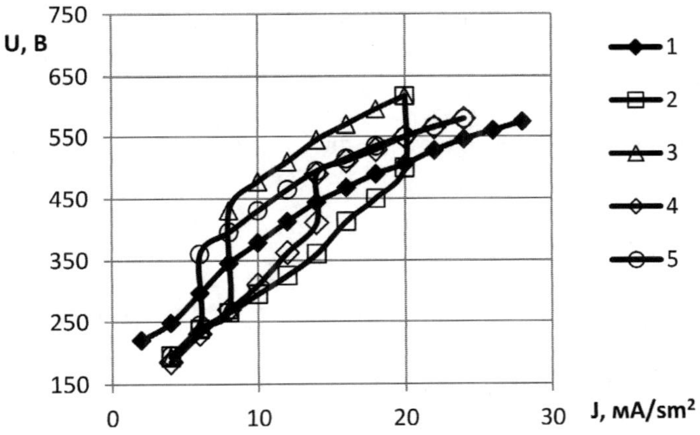

Pic. 3. Volt-ampere characteristics for different work and reactive gas schemas: 1 – pure Argon, 2,3 – Argon and Oxygen mixture schema "1", discharge power increasing (2) and decreasing (3), 4,5 – Argon and Oxygen mixture schema "2", discharge power increasing (4) and decreasing (5)

Oxidation degree of coating at substrate depends on ratio or work and reactive gas in vacuum chamber. Therefore, choosing gas input schema and controlling gas ratio it is possible to manage oxidation degree, which influences on optical and strength characteristics of coating. This method differs from analogical methods described in [13].

Micro hardness (HV) of coatings on PC substrates achieved about 30 kg/mm^2, that is equal to micro hardness of substrate. Coatings of titanium oxide showed 3-rd hardness group according to OST 3-1903-95. Hence this coatings can be used for strengthening of plastic and defend its surface from mechanical wear.

PET substrates with and without coatings mechanical characteristics are showed at pic. 4. Coating that has thickness 50 nm effects on flexure of sample. Thickness increasing enlarges effect. Flexure decreasing means that system "substrate-coating" became harder.

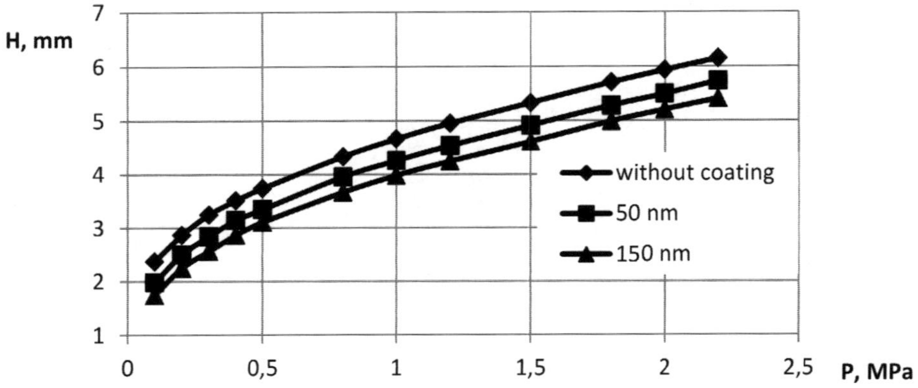

Pic. 4. Experimental flexure H of samples depending on pressure P.

4. Conclusions

Volt-ampere characteristics for different schemes of introduction reactionary and working gases are built. A new method for controlling the level of oxidation coating is proposed. Optimal regime of making oxide coatings on a plastic substrate, taking into consideration the input of gases and low melting temperature of the substrate material, is selected. Micro-hardness of coatings on PC and PET is about 30 kg/mm^2. PET substrates coated TiO_x ($0<x<2$), which thickness up to 150 nm, have 10-30% higher stiffness than PET film without coating. Coatings TiO_x correspond to the 3rd group of abrasion resistance according to OST 3-1901-95 and can be used for surface hardening of plastic. Mechanical characteristics of titanium oxide films provides using this films for strengthening and defending from mechanical wear plastic surface.

References

[1] Galyautdinov R.T., Kashapov N.F., Luchkin G. S. // Welding International. - 2003.- V. 17,- № 8.- P. 655 – 658.
[2] Galyautdinov R.T., Kashapov N.F., Luchkin G. S. // Applied physics. . - 2005.- № 6.- P. 88-93.
[3] Galyautdinov A.R., Galyautdinov R.T., Kashapov N.F., // Bulletin of KGTU. Tupolev. 2010.- № 4. - P. 122-128.
[4] Galyautdinov R.T., Kashapov N.F., Luchkin G. S. // Engineering-physical journal. -2002. - T.75.- № 5. - P 170-173
[5] Luchkin A.G. // Bulletin of KGTU. 2011.- № 16. - P. 121 – 126
[6] Galyautdinov R.T., Luchkin A.G., Luchkin G. S. // Low-temperature plasma in processes of deposition of functional coatings. collection of articles IV Republican scientific-technical conference: collection of articles - Kazan : KSTU Publishing house, - 2013, - P. 155-161
[7] Kashapov N.F., Luchkin A.G., Luchkin G. S. // Bulletin of KGTU. 2010.- № 2. - P. 340 – 345
[8] Luchkin A.G., Luchkin G. S. // Bulletin of KGTU. 2012.- № 15. - P. 208 – 210.
[9] Suikovskaya N.V. // Chemical methods of obtaining of thin transparent film. - Leningrad: Khimiya, 1971. - 200 P.
[10] Yakupov N.M., Nurullin R.G., Yakupov S.N., // Bulletin of machinery building 2009.№6.C.44-47.
[11] Yakupov S.N. // Mechanics of composite materials and structures. , 2010. T.16, N3. C.436-444.
[12] Komlev R.A., Shapovalov V.I., Shutova N.S. // Technical physics letters. 2012. T. 82. Issue 7.- P. 134-136.
[13] Berlin E.V. Ion-plasma processes in thin-film technology/ E.V.Berlin, L.A. Seidman, - M. Technosphere, 2010.- 528 c.

AUTHOR INDEX

Institute of Physics
Dirac House, Temple Back
Bristol BS1 6BE UK

CURRAN ASSOCIATES INC
proceedings.com